Product Managem...
High-Growth Com...

BUILDING ROCKETSHIPS

Oji Udezue AND

Ezinne Udezue

Published by Damn Gravity Media LLC, Chicago

www.damngravity.com

ISBNs:
Hardcover: 978-1-962339-04-9
Ebook: 978-1-962339-05-6
Paperback: 978-1-962339-06-3
Digital Online: 978-1-962339-07-0

Printed in The United States of America

Advance Praise for

BUILDING ROCKETSHIPS

Building Rocketships provides the precise frameworks and strategies that ambitious builders need to create world-changing products. Oji and Ezinne have distilled decades of hard-won wisdom into a practical playbook for building customer-obsessed products that achieve exponential growth.

—Tope Awotona,
Founder and CEO of Calendly

A rich playbook that is packed with practical advice, unique frameworks, and step-by-step guides for building your product-led business.

—Lenny Rachitsky,
Author of *Lenny's Newsletter*
and host of *Lenny's Podcast*

As someone obsessed with PLG metrics and pricing strategy, I found *Building Rocketships* to be refreshingly precise about what to measure and why. This book will become your growth Bible.

—Kyle Poyar,
Co-founder and operating partner, Tremont,
and author of *Growth Unhinged* newsletter

I've seen firsthand how Oji and Ezinne consistently build remarkable products and product teams that deliver real value. I'm thrilled they've packaged their knowledge into a comprehensive playbook for transforming how companies build and ship products. This book provides practical, extensive advice for creating the systems and culture needed for sustainable, customer-driven growth.

—Melissa Perri,
Founder of Product Institute and CPO Accelerator,
author of *Escaping the Build Trap*

CONTENTS

FOREWORD

By Ted Yang, Partner, ProductMind

Product management is not for the fainthearted.

Unlike every other discipline in a company, product management is about what is possible and not what *is*. As a product manager, you create something that doesn't currently exist by synthesizing the spoken and unspoken needs of customers and the market. To make things harder, the practice of product management only exists within and among other peoples' roles. You can execute only through the efforts of people who know more than you about their own areas of expertise. Sounds like fun.

This is why when I first heard about product management I thought it was just a rebrand of project management—that it was all about planning and getting things done. Certainly, this is a huge part of the job, but product management envisions so much more. It is no less than the fundamental operating system of any company.

Ok, great, I said, but where could I go to learn more? As an entrepreneur I was fortunate enough to have the late, great Bob Dorf as a friend and mentor. Bob was a marketing and startup legend everywhere but he started his journey nearby in Stamford, Connecticut, and built his agency there. He was kind enough to be the keynote speaker at our first Startup Weekend back when I was desperately trying to launch the startup ecosystem at the Stamford Innovation Center. Of course, he is best known for *The Startup Owner's Manual*, a fantastic resource for anyone crazy enough to want to build from scratch. That book predates the explosion in popularity of product-led companies, but does contain a few important product-led elements such as customer discovery. Although I have to apologize to Bob that I have never completed one hundred interviews!

Since that book launched, product-led philosophy has spread everywhere and for good reason. After all, isn't building startups all about building products? And therefore doesn't it make sense for companies to be product-led? But to this day there are very few places and very few people who will tell you *what that actually means*. Good luck trying to find a curriculum besides this one written by Oji and Ezinne for training product managers. If you are fortunate enough to know Oji or Ezinne, you could ask them how they overcame various challenges. *Can you tell me what to do with customer feedback? How to price multiple offerings? Or how to make my product viral?*

Now all of their experience has been distilled into one place. In your hands, reader, is a guide that will tell you how to build a product-led organization. What's more, the people doing the telling are two of the most capable practitioners today. There are a lot of product gurus out there, but Ezinne and Oji are walking the walk, everyday guiding companies using their methodologies to be great.

Why has it taken so long? I think because it is easy to throw out ideas and words but much, much harder to implement and make the real-world tradeoffs necessary to truly be product-led. Nothing worthwhile is easy and it helps a lot to learn from the best. Now you have in your hands a roadmap that will get you there. As we say at ProductMind, execution isn't the most important thing, it is the *only* thing.

I first met Oji at a startup event in the Stamford Innovation Center. Despite the fact that we both worked at and were profoundly shaped by our experiences at the Ray Dalio–led hedge fund, Bridgewater, we didn't overlap long enough to interact meaningfully. It took an introduction from a mutual friend for us to discover that we were very similar people who were both hungry to build products and companies. At the time, Oji was working on Mingl, trying to solve a problem with in-person networking at events that still exists today. I was (and am still today) Chairman of a company that hosted many events, and so of course we immediately thought of combining forces. That effort failed.

Why it failed is interesting. The problem Mingl was solving was not sharp enough for my company or the market at large, but more importantly we were not ready to be a product-led organization. We had some visions and ideas but no map to make it real. We spent time and money figuring this out, and even though our companies went separate ways, Oji and I solidified our friendship.

Shortly thereafter, I needed someone to head product at an innovative ad-tech company I had founded. Oji was committed elsewhere, but said that I needed to meet Ezinne. When introducing her for the first time, Oji said (very wisely) that she was "ten times the product manager I am."

My first reaction on meeting her was amazement that two incredibly smart people who were so great at product management could survive under one roof! Ezinne was everything Oji said she was, but I was unable to convince her to sign up. As it turned out, I really could've used her as that company delivered 50–60% improvement but not the 3–5x improvement needed to catch fire. And so I'm one of the few people over the years to try and fail to hire *both* Ezinne and Oji.

Both of them are sharing their accumulated wisdom here, starting from principles and leading to practical how-to steps. Importantly, they are sharing stories that show what happens when things go wrong. Their experiences and where they have practiced their craft differ, which means that they can provide multiple perspectives on the common challenges facing product management professionals today.

Your journey to building a product-led rocketship starts with understanding what the sharpest problems are to solve. If you don't get that right, your rocketship isn't going anywhere. From there, this book covers the most important facets of the product process in detail with an emphasis on how product-led companies do it differently. In my past, I've focused more on companies and technologies than product, and so I know firsthand just how valuable hearing this from an experienced professional is.

Most books would probably stop there, but your company won't truly take off if product is looked at as an island. In a true product-led organization, all of the divisions of the company are working together. Ezinne and Oji introduce the Shipyard and describe how to instantiate one inside your startup or company and make it sing with the whole.

This book is the roadmap, but there is more that we can do to help you on your journey. Our goal is for all of our readers to reach their North Star, and with this in mind we are launching ProductMind, bringing together a family of offerings and services to level up your company. I'm honored to be working together with Ezinne and Oji to bring all of this together.

The Pro Edition of this book gives even more examples of how to and how not to implement a product-led organization. It includes actual artifacts that Oji and Ezinne use for the day-to-day running of their own companies. The ProductMind online community is a great resource for both aspiring and successful product people to learn from each other. The three of us will be offering more content such as talks, videos, courses, and customized services on ProductMind.

As a many-time entrepreneur, I am extremely excited to finally have *Building Rocketships* in my hands. Over the years I've been one of many who have urged Oji and Ezinne to put their knowledge into book form, and while the journey has been long, the final product is fantastic. I know that you will come away from reading it ready and excited to implement its ideas.

Let's build.

INTRODUCTION

World-changing companies are being built faster than ever. The question is, how?

This is the central challenge in product management today and the focus of this book. Milestones that once took decades now take weeks. What is fueling these rapid-growth companies? These rocketships?

Take ChatGPT, the generative AI platform. OpenAI launched ChatGPT without fanfare in November 2022 as a showcase of its latest technology, the GPT 3.0 mode, launched a year before to developers and researchers. In five days, the app amassed over a million users. Then it became the fastest internet service to reach 100 million users in just two months. By contrast, it took TikTok nine months to reach 100 million users, and Instagram a glacial two and a half years. By January 2024, OpenAI had exceeded over $1B in annualized revenue and was valued somewhere north of $30 billion.

In hindsight, OpenAI's success seems inevitable, but it was far from predestined. The startup was originally founded as a non-profit research organization in 2015—the fact they made any money at all was never part of the plan. Despite over $1 billion in pledged funding and access to some of the world's top AI researchers, the organization made little progress for the first two years. Then in 2017, an AI team at Google invented the *transformer*, the neural network–based technology that kicked off the genAI revolution (GPT stands for Generative Pre-trained Transformer). OpenAI created three successive versions of their Large Language Model (LLM) using the GPT technology and saw glimpses of its potential, but they were bottlenecked by the time it spent to train the models on vast amounts of data. LLMs can "learn" with every interaction, refining their responses based on customer feedback. The small OpenAI team couldn't train the models fast enough on their own, so they decided to build a simple but elegant chat interface and launch it to the public—for free. The rest is history—or rather, the future.

Let's look at another example of a rocketship, one we (Oji and Ezinne) are intimately familiar with: Calendly. The SaaS scheduling tool was founded in 2013 by Tope Awotona in Atlanta, and Oji led product at the startup from 2018–2020. In that time Calendly grew from $25 million to nearly $100 million in annual revenue. Incredibly, the company was essentially bootstrapped—Tope raised just $550,000 in the early years—and their technology was not earth-shattering. Critics incessantly predicted Calendly's demise as soon as Google or Microsoft added a scheduling feature (which they did, but each tool worked only with the company's own calendaring services). In 2021, the company raised $350 million on a $4B valuation.

Again, with the benefit of hindsight, it is easy to imagine Calendly's success . . . but why? How did an underfunded startup beat the behemoths of the tech industry who seemed to have every advantage?

The origins of ChatGPT and Calendly could not be more different. One was built on the bleeding edge of AI technology, the other on a mature SaaS platform. One was funded by Silicon Valley billionaires, the other bootstrapped in the Deep South.

Both, however, harnessed and exploited the power of *product-led growth*.

Companies today, both startups and enterprises, feel unceasing pressure to grow faster and more economically than ever before. The winners of this growth arms race are rewarded with higher valuations and market-share dominance, while the losers quickly lose investment and dissolve. The companies in the middle of the pack often face the worst fate. They are doomed to stagnation and the slow bleed of both talent and resources. They eventually lose the war of attrition, wasting years of precious lifeforce in the process.

We have been part of Winners, Losers, and Middlers. We have led teams through rocketship growth and have experienced startup failures firsthand. But our biggest regrets come from the wasted talent and resources at stagnant companies. We wrote *Building Rocketships* to help startups find growth quickly, and for stuck companies looking to reignite the spark to become the rocketship they were destined to be.

This book is for builders, product managers, product leaders, founders, and executives who share our fundamental insight:

> **The most reliable way to create a fast-growing and sustainable company is to build lovable products with a delightful end-to-end customer experience—and the very best framework at our disposal to do this work is Product-Led Growth.**

ROCKET FUEL: THE RISE OF PRODUCT-LED GROWTH

Product-Led Growth, or PLG, is not just a sales or go-to-market strategy. It is an evolution in the way products are invented and businesses are grown.

The core tenet of PLG is **customer obsession** across every product interaction. While this may sound obvious, it is a massive shift from earlier eras of the software industry. Fast-growing companies in the past had to build out massive sales forces and marketing operations. Today's winners, like ChatGPT and Calendly, invest more of that time and money into building the best possible product and customer experience. They know satisfied customers are the ultimate sales and marketing assets, which makes marketing and sales investments more efficient.

It's not just startups that are growing at rocketship rates. Many large enterprises are re-inventing themselves and shipping faster than ever by adopting the core principles in product-led growth. HubSpot, for example, popularized inbound marketing but relied primarily on sales to drive growth. From 2006 to roughly 2014, most HubSpot prospects talked with a salesperson and went through a live demo. It was a high-touch, high-friction process that wasted precious time and resources. HubSpot's head of sales, David Barron[1], eventually pushed for and architected a new approach: Prospects would sign up and demo HubSpot on their own without the assistance of sales. Today, HubSpot's sales organization focuses only on the largest accounts. HubSpot was doing PLG before the term even existed, and the company's value has catapulted since their transformation.

Another example is T-Mobile, where Ezinne worked for ten years as Product Leader in New Products and Innovation. T-Mobile was the fifth largest telecom provider in the US when Ezinne joined, trailing behind giants like Sprint, AT&T, and Verizon. The company recognized that the industry was saturated, and they needed to do something disruptive to break through.

Ezinne was part of a special "Brandstorming Initiative" team formed to tackle this problem. Their mandate was clear: Figure out how to disrupt the industry and propel T-Mobile to the top of the pack. The team started by deeply examining customer pain points. Through extensive research and interviewing, they identified four key customer frustrations: contracts, limited calling minutes, the need to pay for expensive phones, and poor customer service.

Armed with these insights, T-Mobile made a series of bold moves that became known as the UnCarrier Brand Campaign. They became the first telecom provider to eliminate contracts, offering flexible month-to-month plans instead. Then they gave all customers unlimited talk and text. T-Mobile made a significant investment in customer service, selecting call centers in the Philippines where they could deliver better service at a lower cost. Finally, they offered more ways

---

1   Maybray, Bailey. "Product-Led Growth: Examples and Benefits." *Hubspot Blog* (blog), January 19, 2023. https://blog.hubspot.com/service/product-led-growth.

for customers to buy their phones, in some cases simply leasing the devices as part of their monthly payments.

Note that T-Mobile's focus wasn't on a single product, but on *the customer experience as the product*. These initiatives required immense coordination and buy-in across the organization, and the results were dramatic. T-Mobile rapidly gained market share, jumping from fifth place to second place in the span of four years. It was one of the first times Ezinne saw true customer centricity in action. They didn't just pay lip service to it; they made it the core of their strategy and execution. This is the essence of product-led growth.

In the near future, product-led growth will simply be table stakes. It will just be how the best companies build software for B2B, much like how SaaS has become the standard model. Companies that build significant expertise in being product-led can get a strong head start and rack up gains that can compound to defensible competitive advantages in their chosen market. And those that do not will see slowing growth and stagnation.

While product-led growth can benefit all technology companies, we will focus on B2B software because that is our sweet spot as professionals. Successful business software has two main benefits: It relieves toil and provides superpowers to its customers. It should simplify and compress workflows (toil reduction) needed to execute work where possible, or open up new possibilities (superpowers) that were nigh impossible without that kind of software.

This book is split into two stages. Stage 1 is focused on building products with PLG principles. We cover everything from customer discovery to pricing to building durable virality into your products. Stage 1 will answer the question, "What should PLG-minded product managers *do?*"

Stage 2 will shift to key leadership practices in building product-led organizations. This section is written for more senior PMs, product leaders, and founders who aspire to grow their expertise, careers, and their companies quickly. It's also written for CMOs, CTOs, COOs, and CEOs who want to learn how best to work with technology and product leaders who drive a significant portion of their company growth and return on capital. Stage 2 will introduce you to the Shipyard, product systems, managing talent, and setting strategy.

We tried to provide complete details on a wide variety of subjects in this book, but there is just no way to cover everything about product management and PLG in a single volume. We invite you to dig deeper into the topics by reading our blog at ojiudezue.substack.com, which is hosted by our consulting group, ProductMind.

We also created a Pro Edition of this book that includes bonus material, worksheets, templates, and working space to sketch out your own rocketship plans. This edition of the book is available through our friends at Coda. Go deeper and expand your product management skillset by going to productmind.co/brpro.

Rocketship growth is not reserved for only the most well-funded or technologically advanced companies. With the right principles and disciplines, any company can unlock industry-leading growth and define their own destiny.

EXPLORE THE PRO EDITION

STAGE 1

The Rocketship:
Building Products with PLG

The winner in any market is the company that successfully executes on three interlocking layers.

The first is technology: The actual software, hardware, and infrastructure services that form the core of the company. Think about OpenAI's GPT models or calendaring APIs.

The second is product: How the software or services work (form and function) to solve a customer problem or augment workflows with as little friction as possible. This is the customer-facing layer, like ChatGPT's chat interface and Calendly's event scheduling page.

The final layer is the business: How the company delivers on an overall customer experience and creates a monetization model that turns customer satisfaction into profits. Examples include T-Mobile's UnCarrier strategy and Calendly's simple subscription pricing.

Good product managers are catalysts for software companies because they improve the fit of these three layers. They ensure the company finds product-market fit (PMF) by collaborating with marketing, customer support & success, sales, and other roles to massage the *customer experience* to the best that is attainable. These layers are coupled even more tightly inside product-led companies, to the point that the layers are virtually indistinguishable from one another. Think about how the experience of buying an iPhone, using it, and getting support at the Genius Bar all seem to flow together. That is world-class customer experience in action.

So the core question we get asked all the time: What do product managers do, exactly? Here is an incomplete list of responsibilities:

- Perform customer discovery to divine what customers will find real value in, i.e., what people will pay for.

- Collaborate with engineers and designers to build a desirable product that delivers on that value with a view to commercial potential.

- Organize work in the right sequence and scope so that the product can be realized as quickly as possible.

- Help a product team make critical decisions that optimize form and function and time to market.

- Work to ensure that the entire customer experience is delightful beyond just the code and product. This includes considerations like branding, pricing, sales, and support and other services that make up the full customer experience. They may not drive these things but they serve as critical contributors.

- Remove impediments to growth by absorbing ongoing customer feedback and driving customer acquisition, adoption, and evangelism.

A good way to understand the modern product manager is to think about the role in reverse of what people assume product teams do: Product managers convert business strategy and financial goals into a physical product that produces tangible customer outcomes and value.

Many PMs have an impressive array of skills. They must understand technical computer science details (many are engineers in their own right), software testing, data and analytics, customer experience and design, customer interaction science, business model design, marketing, troubleshooting, and more. A PM is a renaissance professional, excelling in many things needed to convert human ingenuity to successful software and tools. It's an expansive role, and there can be many variations driven by industry, kind of company, career seniority, and more.

While the specific skill set may vary from company to company (and sometimes from company division to company division), the fundamentals of the traits and process needed to build highly successful products can be broken down and described, which we will do in Stage 1 of this book.

A product manager is, first and foremost, an expert in customer pain. How do they build that knowledge? That's the subject of chapter 1.

ONE

Customer Discovery: Identifying Sharp Problems

The root of all successful product-led companies is a deep understanding of their customers—particularly of the problems that cause the most pain or friction. We call these "sharp problems." The solution to a sharp problem is often a brand-new way of working—once customers see the light, there's no going back.

As humans solve more and more technology problems and build useful abstractions and refinements for software (e.g., more customer-friendly databases, accessible computer languages, mobile computing, and the internet), what creates value is becoming increasingly subtle, because even what used to be magical software is the new normal. A lot of low-hanging fruit has been long solved in the previous forty years of the PC and internet revolution[2]. What matters now is the final form and function of the software solution and how well it fits the customer's workflow.

We truly believe that even brilliant technology ideas are effectively worthless without connecting them to customer pain. This is even true of AI. *Genius engineering is no longer enough.* Product-led founders and product managers must be connoisseurs of customer pain and workflow inefficiencies.

Improving workflows by a significant factor is key. The job of product management today—especially in B2B software—boils down to *replacing an old imperfect*

---

2   This is not an absolute. There are still a few things that show up as absolutely essential technology layer advances, like better conversational AI or immersive AR that actually works.

workflow with a new, faster, simpler, and more productive workflow. If a founder or PM can accomplish this while delivering workflow compression of greater than 3x, they will earn thousands, maybe even millions, of new customers. If their business model is efficient, it will also make them and their investors and shareholders rich.

Technology companies who make software (or hardware) fail when the "new and better" workflows they create for their target customers are either:

1.   Not sufficiently better than the old workflow

2.   Better than the old workflow, but the problem is not sharp enough for the customer to go through the struggle of changing how they do things (i.e., it doesn't inspire taking on the switching costs)

Without a sharp enough problem to solve, even the most elegant solution will fail. This is why customer discovery (and thus problem discovery) is step one of building rocketship products.

BEHIND THE >= 3X RULE

PMs often ask how we got to the >=3x improvement rule. First, it's not exactly a "rule" because every technology vertical is different. Rather, it's guidance; don't try to build a startup or new product on incremental improvements. For example, it may not be wise to create another AI meeting recorder where the only upgrade is the way meeting summaries are displayed. That is probably not enough to build a large market. If the change is useful, competition can easily replicate it.

Unless you're in certain saturated markets where even incremental gains (less than 3x) are a big deal (e.g., supercomputing), you're unlikely to drive significant adoption. Some key aspect of your product must offer more than incremental benefit.

Peter Thiel, in his book *Zero to One*, recommends the threshold of 10x. We think he's trying to pad the margin of success. In the classic Harvard Business School article *Eager Sellers, Stony Buyers: Understanding the Psychology of New-Product Adoption*, John Gourville's research shows the threshold is somewhere between 3–5x because people tend to overvalue the status quo. 3–5x is the threshold he found where customers consider the switching costs to be worth it. This is what new products have to overcome: the perception of value and switching costs.

In reality the threshold of improvement needed is complex, dependent upon the customers and market. For example, we recently bought solar panels that were 1% more efficient than most of the market. In the world of solar panel efficiency, 1% efficiency is a big deal! A 3x improvement in solar panel efficiency is not even technologically possible today, and whoever achieves this will likely win a Nobel Prize.

Overall, in the B2B software market >= 3x is a good target to ensure your business does not struggle and can maintain enough of a lead to gain revenues to fund further advantages.

DEFINING THE THRESHOLD OF SHARPNESS

Sharp problems is a product framework for first determining if the central workflow problem your company is trying to improve will generate sufficient demand for a successful company. In addition, it can be used for prioritizing what to build and when. Once the sharpest problem is identified and solved, your roadmap should be prioritized to solve the *next* sharpest problem, ad infinitum. Because sharp problems have a high correlation to opportunity, your company will thus always be solving the *next highest opportunity problem*. This gives your product the best chance to successfully replace an old, suboptimal workflow with a new, productive, *can't-live-without-it*–type augmentation.

There are three key questions that need to be answered to identify sharp problems:

1. **What are the current alternatives to solving this problem?**
   How are people "duct-taping" solutions together currently?

2. **How prevalent is the problem?** Do millions of customers or thousands of businesses face it? There are four salient parts to the prevalence question:

   a. **What is the market size of the problem?** How many people or companies on the planet experience the problem and/or how much is it worth to them?

   b. **Is the problem growing?** In the future, will the problem grow or shrink, given current trends?

   c. **Is the problem general or niche?** Does everyone at work experience this problem frequently? Or just a specific role or department?

   d. **Is the problem high frequency?** Does the customer experience the problem daily, weekly, or monthly?

3. **What is the potential "value" of solving this problem?** In other words, is this a problem that people (or their managers) would be willing to pay to solve? To really understand value deeply, it important to understand a related question:

   a. **What parts of the problems are still hard for the target customer?** What saps their time, productivity, and energy? What represents toil?

Strong, convincing answers to these questions will compel the customer take on the switching costs (in addition to your price) to buy your technology product or service.

Let's consider a classic *sharp problem*, one that Oji was deeply familiar with: *scheduling meeting times with customers who are not in your company*. This also happens to be the raison d'être of Calendly.com, a company that Oji worked at as Chief Product Officer.

CALENDLY'S SHARP PROBLEM

Since the beginning of recorded history, humans have always needed to meet to accomplish things together, be it work-related, political, or personal. For many modern-day professionals, particularly those in sales, marketing, and recruiting, you need to meet with partners, prospects, and candidates. Theproblem is that there is no real way to know their schedules ahead of time.

Therefore, scheduling a meeting comes with some kind of communication to coordinate the participants. Once you add distance and time, coordination can get quite complicated. To get a single meeting you usually have to exchange multiple messages, which can be tedious. It's much easier to do with people who are close to you and in the same village or directory.

Usually your workflow starts with a good outreach to a potential customer, which engenders interest. Now you want to meet to make a connection or a sale. You then message the person to inquire about their availability and share yours. This can happen over email or text messages. You go back and forth a few rounds; this can be longer if your schedules have no initial overlap in the near term. After a few rounds, you land on a good time with sufficient urgency for both of you, and then you meet.

This outlined workflow can take a few minutes to a few days. Often it does not converge at all, leading to no meeting, because it's hard to manage many of these at once. The more of them you have to negotiate, the more meetings fall through.

As previously mentioned, almost everyone on the planet goes through a version of this, but professionals who need to do this at high volume to make their goals (sales numbers) and their livelihood need as much assistance as they can get. Thus, they will likely *feel* the workflow assistance from an effective solution.

DIMENSIONS OF CALENDLY'S SHARP PROBLEM

By answering the questions from the previous section, we can see that the problem of meeting scheduling is sufficiently sharp:

- Widespread: It's experienced by almost everyone on the planet, especially if the recipient is not in your directory. Virtually everyone has experienced the pain of coordinating schedules.

- High-Frequency: The toil factor increases linearly with how frequently you have to do this, and for some professionals, it can be quite frequent (e.g., salespeople, recruiters, and marketing pros). As the frequency increases, error rates increase.
- Growing Problem: It's being exacerbated by modern life, with no end in sight. We're busier than ever, and there are more connections to be made on average to earn a dollar. More and more business is being conducted over the internet and from almost every location and time zone in a globalized world.
- Value: When a meeting is not booked, the initiator may lose sales or potential new business, and the recipient may lose a valuable opportunity.
- Alternatives: There are no easy alternatives at high-frequency scale of operation or role.
- Willing Buyer: Managers and supervisors of certain kinds of professionals deeply care about this problem because solving it can have a measurable impact on the bottom line.

Thinking through the criteria makes it quite clear—Calendly is solving a sharp problem!

Let's discuss the key questions below:

What Are the Alternatives?

Thinking about alternatives is an exceptionally great way to find and assess sharp problems. What problem has no workaround? If a customer can find a trivial alternate solution to the workflow problem you're trying to solve, it's NOT a sharp problem. At Calendly, for example, customers wanted to check their multiple calendars for conflicts before scheduling a meeting. Early on, Calendly could only talk to one calendar at a time. The Calendly team's research showed that most customers had at least two or more calendars, and there was no viable workaround other than managing two separate Calendly accounts (which a few power users did, but it was extremely cumbersome and hardly solved the coordination problem). When prioritizing features, connecting to multiple calendars was a sharp problem worth solving for some power customers.

On the other hand, Oji's team chose not to ship a feature that seemed obvious to many users: the ability to email a Calendly link from within the Calendly app without using your email application. There was a viable alternative that Calendly customers still use to this day, which is copying and pasting their scheduling link into email or text. Sharing your link is not a sharp problem.

By the way, this didn't mean the feature wasn't useful. Oji's team planned to pair the email feature with a different sharp problem solution that could only be solved by building this first: automatic scheduling reminders sent to a contact if they forget to schedule a meeting after the link was sent. Calendly customers hated following up with contacts, so this was a legitimate sharp problem.

Sharp problems are not always obvious. They can typically only be discovered via customer immersion and knowledge. The best PMs and founders tend to be those who are *deeply and personally insightful* about a problem that needs solving. The best of the best identify problems felt by millions of people or thousands of companies.[3]

How Prevalent Is the Problem?

The prevalence of a problem can be measured by four vectors:

---

3   Some problems don't apply to a lot of people, because those who experience them have an atypical workflow or environment or are singletons in the economy. Even if the solution meets the rest of the criteria, avoid these. You need scale of target customers to benefit from the economies of scale in software businesses. Otherwise, you have to charge each customer a lot of money to compensate.

1. Market size: Is this problem felt by billions of customers, hundreds of thousands of customers, or thousands of enterprises?

2. Problem growth: Is the problem growing as the world advances? Or will another technology solve it in an indirect way that will reduce the problem's footprint over time?

3. Nicheness: Problems that are felt across the enterprise are sharper (and more valuable) compared to problems isolated in a specific niche, role, or department.

4. Frequency: How often do customers use tools to solve the problems in their workflow? The range is usually rare, yearly, monthly, weekly, daily, or multiple times per day.

Product managers (and founders) need to consider how many people are impacted by the problem. Is the problem felt all over the world by many people and companies? If the problem is not, you must carefully consider the value of solving that problem since you will have fewer customers to sell to. Note that you can build a large business by selling to a hundred Fortune 1000 enterprises but likely not to a hundred ultimate frisbee players.

The next vector is the growth rate of the problem. Ideally, you want to solve a problem that is only growing or getting worse. Google's growth depends on ever-increasing internet penetration and the proliferation of online information in need of organizing. Basically, as the population and the internet grows, the need for Google Search increases.

Calendly's success was predicated on the fact that more and more meetings are occurring with people outside their organizations and most of them are virtual— this was true even before the 2020 pandemic. So ask yourself: What trends and variables drive my business forward, and are they up and to the right?

Nicheness is an important idea to grasp in determining prevalence. Is this a broad-based problem, felt by many people across industries and professions, and even in personal life? Or is this problem confined to a specific market of customers? A niche problem can still be global and large, but the go-to-market plan and business model will look very different from that of a general problem. A problem felt by most workers or people can have much larger markets everywhere.

Finally, the more frequently a customer encounters a problem, the more valuable your solution could be to them. For frequent tasks like scheduling meetings, small increases in productivity matter a lot. Similarly, early winners of the Gen AI evolution have focused on marketers, copywriters, designers, and software developers. Their tasks are frequent and AI can improve their workflows by up to 3x or more. They are usually willing to pay for well-implemented software technology to help reduce their toil and increase their output.

Another way to think about frequency is to additionally consider how long a specific problem space has been around. If you consider our most foundational needs/workflows as humans, you'll typically find a massively successful technology business. For example, acquiring goods and resources (Amazon), communication (WhatsApp, Facebook), information gathering (Google, ChatGPT), group coordination (Calendly, Outlook), accounting (Excel), selling (Salesforce), and more. In other words, the older the problem, given high-frequency, the sharper it likely is.

What Is the "Value" of Solving the Problem?

There are many sharp problems in the world that, unfortunately, don't equate to business value. This is the domain of governments or philanthropic work that aims to improve the world for its own sake, which is a noble endeavor, but outside the scope of this book. Speaking strictly of business opportunities, founders and product managers need to identify problems that carry a monetary value.

There are a couple ways to estimate the "value" of solving a problem. The first and most reliable is to determine how much customers are spending on alternatives to solve the problem today. That alternative may be another technology product, a manual task, or an employee or freelancer. If your customers already spend money on solving the problem today, you are on the right track.

Some sharp problems have no budget. Either a solution doesn't exist or your customer isn't aware of the problem. Defining the "value" of solving the problem is more difficult in these situations, but one way is to measure value in time spent. How much is your customer's time worth, and how much could you help them save? This was how Calendly made their initial calculation because virtually no one paid for a scheduling tool, but the problem of scheduling coordination was frequent enough that it cost professionals (and their employers) real money.

ROADMAPS: A SEQUENTIAL LIST OF YOUR SHARPEST PROBLEMS

Roadmaps are often viewed as a list of features to be built, ranked in order of priority by some ambiguous means. Every product manager has seen a roadmap like this, and they have likely created a few in their day.

This is common because ultimately, engineers build things, and part of the product manager's or founder's job is to translate all the business and product gobbledygook into a representation they understand, so they can actually do the building.

The issue with a feature-driven roadmap is that it puts the emphasis on the feature, when in reality, a product team's focus should be 100% on the customer problem or opportunity. The feature should be treated as an *attempt* at solving it with an attendant probability of success measured by well-considered metrics of success. Good product teams don't treat the attempt as the solution unless they can measure its success. Good product teams achieve a high attempt-solve rate.

Instead, product teams and companies should create roadmaps that solve the sharpest problems for their customers. To really understand value well, additional questions are: What alleviates the hardest stretch of the customer workflow goals? What reduces toil the most, and what gives them back time the most? What gives them the most (apparent) superpowers?

These problems or opportunities should be prioritized based on your evaluation of how sharp the problem/opportunity is.

The traditional impact scale, in this case, means the intersection of problem sharpness and the ease/time in which it can be solved. In fact, a product team or company could plot problems on a 2 x 2 graph where the x-axis is problem sharpness and the y-axis is the ease of solving[4].

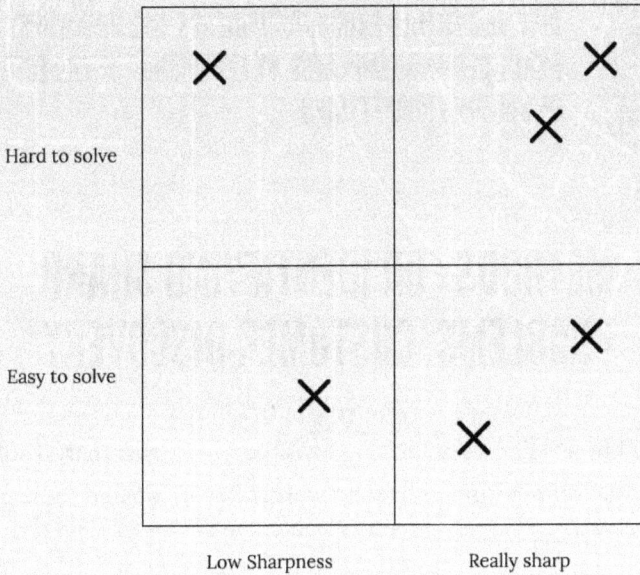

The problems that land in the bottom right (high sharpness, easy to solve) should be prioritized first. Problems in the bottom left (low sharpness, easy to solve) should only be released along with a feature or product that solves a sharper pain (remember Calendly's email link feature?). Problems in the top right (high pain, hard to solve), should not be ignored, but be part of your long-term strategy in the product portfolio. They often offer the most opportunity, precisely because they are hard to solve and create the most feature moat. Problems in the top left (low pain, hard to solve) should usually be scrapped from the roadmap.

Often multiple solutions or features will be needed to address a sharp problem comprehensively. It gets even more complicated when they have to be delivered by different teams with different skill sets (this is very common). In this case, it's advantageous to isolate the hardest part of the holistic solution and solve it, in order to create a foundation for stacking the other parts of the complete solution to the problem—this is akin to breaking down the sharp problem and stacking its pieces.

---

4   Note that ease of solving is not a technical measure alone. A product is
     both the technical layer and the go-to-market. Therefore ease of solving
     includes how easy it is to introduce to the largest number of customers.

In short, our advice is: Solve the sharpest problem, especially the ones that let you deliver quickly. If it can be broken down into several pieces, build from the hardest piece to the easiest piece.

DIVE DEEPER INTO PRODUCT ROADMAPS, INCLUDING BEST PRACTICES AND TEMPLATES, BY EXPLORING THE PRO EDITION OF BUILDING ROCKETSHIPS

METHODS FOR IDENTIFYING SHARP PROBLEMS: CUSTOMER DISCOVERY

So now we know what we're looking for: sharp problems that, if solved, will change the way your customers live and work. How do we find those problems? That's *customer discovery*.

Customer discovery is the deliberate practice of studying the workflow patterns of your typical customer to understand their motivations, goals, and frustrations in order to improve and streamline it using technology[5]. When this study is done well, a researcher (usually a product manager or designer but also often a trained user researcher) will observe key pain points, but also opportunities to delight and amaze a customer subject. They will uncover the sharpest problems to solve for that customer.

Let's talk specifically about discovering workflows, then turn our attention to customer discovery techniques.

---

5   Technology in this case is usually a software tool in the workplace. But it
      can also be hardware or a software-hardware combo.

Workflows: The Source of All (Your Customers') Problems

Workflows are the sequences through which people get tasks, projects, or their jobs done. For any given role at work, there are a million workflow variations for the same role in businesses worldwide. Good technology (hopefully from your company) will try to help customers through a better workflow sequence, so they can have the most efficient and delightful experience possible[6]. These technology-enabled workflows allow customers to complete tasks more quickly, with less hassle, and with higher quality. Any *magical* moment you've experienced with technology was the result of discovering a radically better workflow using the new technology you acquired.

When Marc Andreessen declared that "Software is eating the world," what he really meant was that we are in an era of replacing analog or manual workflows with software-based workflows. Virtually every workflow can be continuously improved by better and better technology. For example, even the best-run taxi services struggled to operate as efficiently as Uber, despite having decades of experience over the startup. Taxi technology was dated and ignored clearly painful customer problems like hailing a cab, haggling over price, cleanliness, etc.

The role of the product manager, then, is to discover not just parts of their customer workflows that can be dramatically improved, but opportunities to improve or reimagine them with technology.

B2B software, in particular, is incredibly profitable because corporations are willing to spend a lot of money to be more productive and efficient. Instead of companies spending time to discover the most efficient workflow for every business function, they outsource that work to trusted software vendors who have seen incalculably more variations of the workflow, have identified what works best, and are using technology to implement the best ways of accomplishing a specific job.

---

6   Most people do things a certain way for many reasons, including history and culture. Few know the truly "best" way or workflow. Good software can herd people into the top workflows that will help them succeed.

Workflows, however, are affected by motivations and environment. Context-sensitive considerations include: Why does the customer need to do this particular task, and for whom? Who else is on their team, and what are *their* preferred workflows (and are they the same)? What sort of support (both leadership and resources) does the customer have to choose their optimal workflow? How much autonomy do they have?

The process of customer discovery must uncover how customers *actually* work in order to identify issues with the workflow and prioritize the sharpest problems—not an idealized dream of how they or a researcher *thinks* the customer works. Only with a focus on the actual workflow can a product manager apply technology to make things particularly better.

We recommend that the "researcher" shouldn't keep their workflow discovery too narrow. A good customer discovery process inspects three distinct but related workflows: the target workflow, the preceding workflow, and the succeeding workflow.

The target workflow is the one your product is trying to optimize. This is the main *thing* your feature, product, or company does. Luckily most technology companies have a clear idea of how they reduce toil and augment the work experience for their target customers. The preceding workflows are the activities that the customer is engaged in right before your target workflow. The succeeding workflows are the things they do immediately following your target workflow. This can be referred to as the workflow *value chain* (a nod to a common term used in analyzing business strategy).

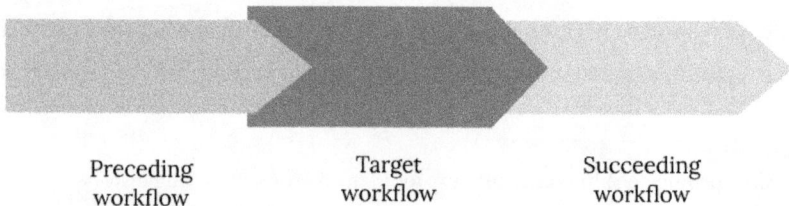

| Preceding workflow | Target workflow | Succeeding workflow |

It's important to understand not only the motivations and environments for these secondary workflows, but also how employee-made outputs flow through all three stages. Deeply understanding the pre- and post-target workflows also gives a product team or company a clear *value expansion plan* in the future for their customers by augmenting *more* of their everyday work. Most software becomes more capable over time by eventually absorbing preceding and succeeding workflows.

Workflows are the core of good customer discovery, particularly for B2B and "prosumer" problems. Here are eight more principles to improve your customer discovery process.

Customer Discovery Is Continuous

Henry Ford famously said, "If I asked customers what they wanted, they would have said a faster horse."

While customer discovery often involves customers narrating their experiences, this narration should never be taken at face value, especially when customers are very opinionated about what they need. Product teams should not completely rely on customer discovery to tell them *how* to solve a sharp problem. The first priority is to really ascertain that the problem exists and is sufficiently painful to be worth solving. Only then is it appropriate to draw inspiration for the customer narrative[7] about solutions.

Product teams must look for unspoken insights that give clues to motivation, environmental constraints, key stakeholders, toil, and more. Customers, like all humans, are not necessarily aware of the stew of things that drive their actions, although they do tend to be more perceptive of this in the work context.

---

7   Some customers are very insightful about solutions, but not all. Listening and discernment is important.

Customer discovery is not static or finite. One research study is never enough. Instead, what is needed is *continuous* customer discovery. Product-led companies need to invest in the ability to continually study target customers for insights into what matters most to them. Continuous study yields two valuable things:

1. A better ongoing understanding of the customer journey and any changing needs.

2. The opportunity to correct inaccurate conclusions about the customer journey.

Remember that a single customer conversation is rarely valuable. However, a hundred, averaged together, will almost always yield real insight that can be combined with the creative process to produce technology that solves a real problem of value.

Customer Insights Must Be Timely

The best teams work hard to deliver on speedy but accurate discovery. Then they share these insights with their organization so that "customer sense" (the organization's intuition about their customer and problem space) can pervade the software development process.

CUSTOMER SENSE

Product managers talk a lot about product sense. In product management, "product sense" refers to an intuitive understanding of what makes a product successful. It involves a blend of insight, instinct, and knowledge that helps product managers make decisions about the features, design, and overall direction of a product. People with strong product sense are able to balance customer needs, business goals, and technical feasibility to build products that provide value and resonate with customers. This book is about honing some of your product sense.

However, we think of **"customer sense"** as a powerful prerequisite for product sense. This is the intuition you get from immersing yourself in customer problems deeply, large and small. Eventually you get a sense of what will work and what will not for proposed solutions. If you want to really understand customer sense, hang out with the smartest customer support rep you know. They likely have it.

When we discuss the Shipyard in later chapters, you will see how we organize a product team to infuse customer sense from the support teams.

Customer discovery can take time—it includes scheduling, iterative interviewing, synthesizing, and dissemination. So it requires optimization as your team scales. Software teams can turn around critical discovery activity in a week or two if they know what they're doing. Here is a three-pronged iterative process for doing effective customer discovery that is possible to execute inside a week:

1. The first step is to select 5–10 representative target customers[8] and schedule them for 30–45–minute meetings. The purpose of this first step is to discover what questions to ask, to understand your target workflow. In these meetings, we ask exclusively environment and motivational questions:

    a. Their title and position in the company

    b. What they were hired for

    c. Who sets their agenda

    d. Their daily workflows and what a typical day looks like

    e. How payments are made for current software

    f. What creates toil and pain

---

8   Always pick customers who are representative of your customer base and target in order to make the interviews valid and a good learning experience. Try to be very specific about the dimensions of the customer to select. Being general degrades the learnings.

These kinds of conversations can feel loose and unstructured. That's totally fine. The researcher should generally have a partner to take notes or a transcribing recorder software (gain permission first, of course!). Often it's a team sport, with engineering, design, and marketing listening in. This first step yields more specific questions to ask in step two, and about areas of interest for the target customer.

Total time elapsed for interviews: 6–8 hours
Planning time: 2 hours
Synthesis time: 2 hours

2. Next, we select 5–10 new representative target customers for another round of 30–45–minute interviews. In these conversations, use the output of the first step to ask more structured questions that pinpoint target workflows and should yield suggestions of the key problems to solve. For example:

    - Can you share your workflow for achieving [this task]?
    - Can you dive more deeply into aspects of how you accomplish this workflow?
    - Can you share with me why you choose this path versus that path, this tool versus some other tool?
    - What kind of tools do you use for this now? What works and what's not working?
    - After this workflow, can you tell me what typically comes next? How do they connect?

    Total time elapsed for interviews: 6–8 hours
    Planning time: 2 hours
    Synthesis time: 10–12 hours

The questions are consistent between customers so it's easy to understand variations and see if areas of inquiry are on-target or off base. This should yield a sense of parts of the workflow that can be optimized profitably. You can choose to come up with some concepts for improvements to the workflow and do a check-in with more customers. If you choose to go more quantitative, note that it comes with its own challenges. However, a third step could look like this:

3. Select 20–80 representative target customers and send them a set of even more refined questions (usually no more than ten) as an online survey. For example:

    - Name
    - Title
    - Company
    - Title
    - Role
    - Scale of pain in a section of the workflow
    - List of potential manager titles who set their goals and agenda
    - Frequency of the workflow in their day or week
    - Types of close collaborators
    - Proposed solution or concept #1, #2, #3
    - Willingness to pay for any of the proposals
    - Whether they need approval to do so

    Total time elapsed for surveys: 2 hours
    Planning time: 2 hours
    Synthesis time: 2 hours

This can quickly turn your qualitative insights into quantitative insights, which can help you focus on the right parts of the problems to solve.

STAGE 1

Discover good workflow questions for target customer

5–10 customers
10–12 hours

↓

STAGE 2

Analyze target workflow and workflow value chain for sharp problems

5–10 customers
11–14 hours

↓

STAGE 3

Optional: Turn your insights into quantifiable data

5–10 customers
~6 hours

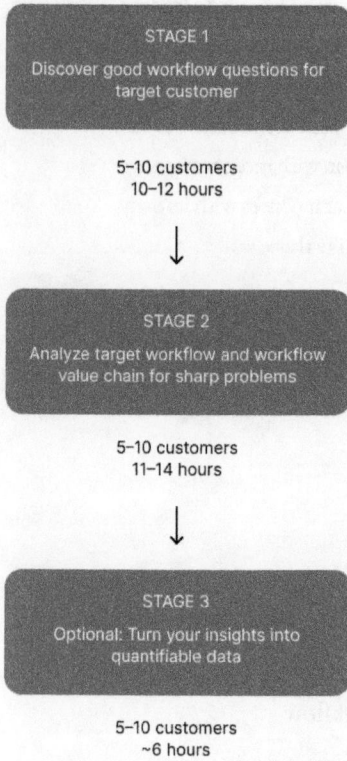

Customer Discovery Is not Just the Job of Researchers

When a software business is young, it's common for successful founders and their first few employees to spend significant time with their customers to try to understand what they need. It's all hands on deck, so even engineers get customer face time. As success comes and scale happens, this connection to customers starts to sever. Usually, it's because people get busier and specialization happens. Soon it's the job of PMs and user researchers to produce insight because engineer time is too precious to waste on that activity. In really large organizations, PMs eventually get out of the job of creating their own insights as well and outsource it completely to customer researchers, data analysts, and customer support. They spend all their time coordinating the myriad teams they need to deliver on ever more difficult projects.

It makes sense why this trend happens. It's expected, even. The job of leaders and founders in a product-led environment is to work to reverse this trend, because what is lost is tremendous—a nuanced sense of what is important to their customer by professionals who are tasked to create products that solve their problems (in other words, customer sense).

A product team has to make thousands of small decisions about what to prioritize and build over a product lifecycle. These decisions are often a balancing act between customer needs and business needs. That balancing act works better when that customer sense is in full bloom in multiple people on the team or squad. However, what typically happens is that only one person (the UX researcher or PM) has sufficient insight to strike the right balance. No executive wants the viability of their business hinging on a single person.

It's critically important to the longevity of your company that product managers are required to maintain their customer sense by spending 5–10% of their time doing continuous customer discovery and customer listening. The very best product teams will require key product team participants like designers and engineering leads to participate in customer discovery as well.

Customer discovery should be considered a team activity—even as the company scales. *Especially* as it scales. Everyone doing creative research and product development should be trained in the fundamentals of continuous customer discovery. When hiring product managers and designers, specifically seek out customer discovery skills and customer sense.

Lower the Barriers of Doing Discovery

Part of the reason that most product teams trend away from customer discovery is that it's hard work. Not only do teams have to interview customers, they have to synthesize those interviews and pluck out insights that help the creative process. It's a multi-step process, mostly analog, and fraught with risk of getting it wrong; thankfully AI helps with synthesis now.

Good product-led organizations will work to lower the barriers for their product teams to do customer discovery. There are a range of tactical things you can do to make it easier. Here are a few suggestions:

- **Routinely ask customers to opt-in to research activities.** Use your software to ask customers if they want to be part of the product improvement process, which might mean occasional interviews by the product team. Make it an opt-in in the UI or via email. If necessary, dangle a reward like an Amazon gift card or something similar. Over time this creates a pool of people who could potentially be a good subject for discovery.

- **Automatically schedule customer conversations with your PMs and designers.** This is fairly aggressive but works wonders when done properly. Rig a system that picks a customer at random and asks them for a scheduled thirty-minute interview with a designer or PM (multiple team members can join a customer call). This should just show up on the PMs calendar one to three times a week. These kinds of ambient discovery calls add up to customer sense, assuming the PM conducting it is trained.

- **Have customer research panels for major feature areas.** Consider setting up standing panels of customers who can be called on more purposely for research. Some companies or teams have a more formal customer advisory board (CAB), which is a great thing. But usually, a CAB is more about providing input on strategic direction setting. What product teams need can be a smaller, more informal version that is more selective of the customers being targeted. We've seen teams form product councils for their product areas.

When Oji was Head of Product for Creation & Conversation at Twitter (now X), the Conversations team had multiple such panels that represented key audiences for us: a set of younger millennials, a panel filled with women—especially those of color—and another for Black creatives who drive a lot of conversations on the platform. But these were only the most formal ones driven by the research team. Major projects created their own groups of early adopters for more intimate research—usually on discord, Slack, or some other groupware. PMs could then ask questions as needed in an organic way and use this to drive insight.

Make Space for Customer Discovery

Another reason continuous customer discovery doesn't happen is that managers often don't create space and time for their product teams to do it. The implication of being continuous means that it is a permanent tax on productive time. This is why we encourage product and company leaders to make it ok for PMs and other product leads to spend up to 20% of their time on this activity.

Whenever we are working on a product, we are almost always constantly doing customer discovery. If we find friends or strangers who fit the target persona of our products, we start to ask questions that help further our insights into their workflows and motivation. For us, airports are a particularly fertile place to do this because we encounter so many random professionals with interesting backgrounds and some perspectives we had not considered.

Create a Customer Discovery Culture

The best product-led companies don't just do customer discovery—it's baked into their culture.

When hiring product managers and designers, they specifically seek out customer discovery skills and customer sense. They celebrate acts and outcomes that resulted from great discovery practices. They incentivize all their teams to become good at discovery and make it a part of their everyday mode of operating.

If at all possible, mandate dogfooding

Another area ignored by many good product teams is making sure that the entire team "dogfoods" consistently. Dogfooding is the act of a) making it easy for the product team to use early in-progress versions of the product, and b) making sure they actually use it to a threshold of frequency that simulates customer use and thus helps team members generate their own insights and epiphanies about the product themselves.

Ezinne's time at T-Mobile illustrates this well. Around 2003–2004, her team was responsible for the handset experience. Months before a new handset would launch from a manufacturer, they would get samples and dogfood it relentlessly. They would critique each menu, each click, working hard to minimize the cognitive overhead of using any clunky interface that was engineering driven. They would spend time with customers and watch them try to accomplish common tasks on these new phones and find all the frictions. You would not believe how hard early mobile phones were to use in the days before Android and iOS hegemony. Every phone had its own OS and menu system. What resulted was a ton of feedback to the software teams of manufacturers to change before the phone got approved on the network. This resulted in massive improvements in the user experience when new phones launched.

A team that dogfoods will help its members understand the frictions implied in the product use and shadow the customer journey. Dogfooding is also an easy way to adjust the product based on early feedback and customer-generated bugs.

One cool variation on dogfooding is taking early concepts or even versions of the product and sharing it with another team in the company not involved in building it. For example, checking in with the customer support or the finance team to get their unvarnished feedback. Their distance from the building of the product was often a fresh perspective that was materially useful in making it better before exposing it to all customers.

All dogfooding qualifies as the *confirmation stage* of customer discovery—figuring out if your translation of the problems into solutions is hitting the mark. It's another valuable way for your team to develop customer sense.

Invite experts in when needed

Since customer discovery is about understanding and streamlining customer workflows, it's sometimes helpful to bring in expert opinions. As VP of Product at Procore, Ezinne's team organized an "Innovation Day" that invited dozens of construction experts to the company to discuss the latest innovations in their fields. Construction is a highly specialized industry, so Procore relied on a community of experts to help steer their product direction.

Be sure to use the time of experts wisely. Don't call them in for mere concept testing or to sit on a research panel. Ask them to weigh in on your product plans as early as possible to ensure you understand the problem deeply enough and your solution is viable. Experts can be an invaluable way to condense insights if you can find the right partners.

LIMITATIONS OF CUSTOMER DISCOVERY

It's now very fashionable to advocate for customer discovery because technology companies *rarely do enough of it*. However, it's also important to understand the limitations. Customer discovery could also be called *problem discovery*, but we do not consider it *solution discovery*. Customer discovery helps product teams answer a key question: Which problem is most important to solve?

Once you understand and prioritize the sharp problems in your customers' lives, product managers have to translate those problems into simple, loveable, and intuitive solutions. This is why design is such a critical aspect of problem management and why we dedicate chapter 3 to the subject.

Before that, however, let's focus on one more critical skill related to crafting and refining the right solution for the customer: customer listening.

TWO

Build a Customer Listening Machine: Capturing and Synthesizing Feedback

Customer discovery helps you understand your customer and uncover sharp problems worth solving. *Customer listening* helps you refine the solution. It finds places of opportunity in your current product, helps you prioritize the key building blocks of your software roadmap, and confirms whether or not the new things you're building are working.

The moment the product team ships a release, their customers start to have a conversation with the product, whether or not it's visible to the company. Customers will cycle through the joys, elations, frustrations and suggestions about the software—but maybe 1–10% will be explicitly directed at the company through formal communication channels. Some of the feedback will go through support and the company's help docs. Most of the conversation is happening on social media and other channels outside the company's control.

A good product-led product management team will work to collect these kinds of communications and do three things:

1. Sift through them to refine their sense of the next sharpest problems (these can lead to bug fixes or entirely new functionality).

2. Use them as an extra source of data to confirm or disprove their synthesis from customer discovery.

3. Look for opportunities to respond to customers when it matters (it's hard to do this for all feedback, so the team needs to discern which responses are most important).

Customer listening is useful for feedback on the products and features a company has already delivered. Sometimes it gives hints on bigger problems that need to be solved (which should be the subject of further customer discovery to really understand it). Customer discovery and customer listening create a virtuous loop that surfaces sharp problems, suggests new workflows, and refines those workflows to make end customers more productive.

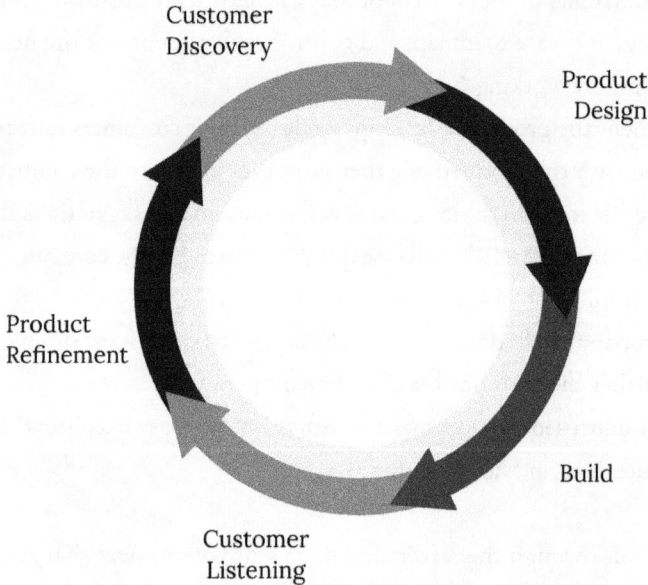

Customer Discovery

Product Design

Build

Customer Listening

Product Refinement

If you're a fortunate product team, you have an upward curve of adoption and your customer base is growing. Soon that 1–10% feedback swells to an unmanageable extent. Your customer feedback becomes difficult to sift through. A customer listening system is meant to help you find the gems within the avalanche of data you're going to get and will help you prioritize to an extent that other competitors cannot.

Here are a few ways to get this right.

LOWERING THE BARRIERS TO FEEDBACK

Oji had the opportunity to work at Microsoft early in his career where he saw the benefits of continuous feedback firsthand. The early adopters of Windows were developers, and that meant Microsoft got a lot of bug reports. This feedback from highly educated customers was invaluable. As Windows expanded to business customers and consumers, Microsoft needed a way to continue getting bug reports, so they devised a plan for Windows software to file its own feedback automatically. If some app or process crashed, an automated tool called Watson (cool name!) sent a "crash dump" back to Microsoft. There, the crash dumps were directed to teams who owned that specific part and then debugged by developers who worked on those teams.

These crash dumps weren't exactly customer "conversations," but they were a form of direct feedback from the customer's computers. For example, if Oji's team suddenly saw the same crash a million times, it was an obvious clue something broke in the latest release. Microsoft could even automatically send crash reports to app developers when their apps crashed on Windows. Today most software teams no longer need Windows or Mac OS X crash reports because the main delivery mechanisms for applications are web or mobile app stores, but this experience reveals a valuable lesson: Companies need to lower the barriers to feedback, even to the point where the feedback is automated. Else you will miss critical information to improve your product.

Not all customer feedback can be automated. In these cases it's even more important to lower the barriers for customers. The product team must make it so easy to provide feedback that anyone who is even the *slightest* bit inclined will do so.

Lowering the barrier to giving feedback is important for two reasons:

1. You will get more feedback to improve the product.

2. The feedback you get will be more representative.

Most feedback is from either the most delighted or the most disgruntled customers, a statistical phenomenon called bipolar distribution. In practice, it usually skews toward the disgruntled side. We live in the age of convenience and people expect software to just work. They don't leave feedback unless there is an outlier experience (usually negative) that disrupts their expectations. The easier it is to share feedback, the more likely you are to hear from customers across the broader spectrum of satisfaction.

So how can you encourage more feedback? Here are some suggestions.

Embed Feedback in Your App

The opportunity to give feedback should be embedded in your application, not hidden three layers deep in the Contact Us menu.

Customers should be able to go to a place in your app to submit feedback whenever they like. It should be a consistent, well-designed experience. The form itself should be structured so it's more likely to yield useful feedback. For example, create categories of feedback so it's easier to filter and sort when you get it on the other side. Make the copy and design inviting and not overly corporate. If your app has a reward system of some sort, tie rewards to feedback, or perhaps offer a free product training session.

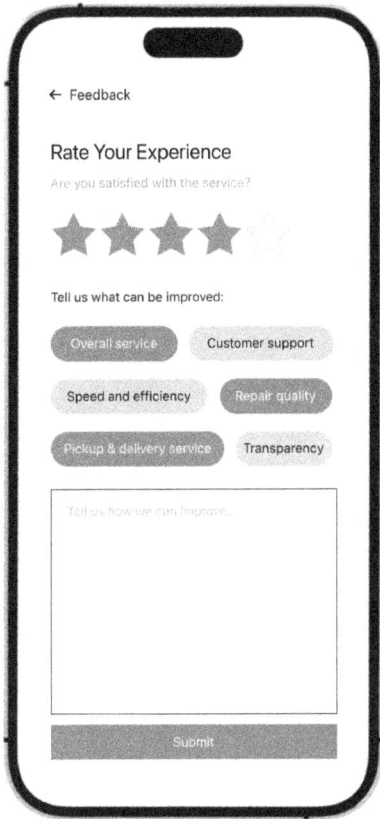

Feedback systems in apps should be *progressive*, so that customers can provide a little feedback or a lot. Feedback can start with a simple question like, "How are you enjoying your experience so far?" and offer a simple thumbs-up or thumbs-down response. Then those who are motivated can be prompted for longer-form feedback. This serves two purposes: It allows low-engagement customers to still weigh in (thumbs-up or down) and more expressive customers to give you rich feedback.

Thumbs-up and thumbs-down systems are a basic form of sentiment analysis and are more useful at the feature level than at a whole-product level. It should always be paired with the opportunity to share more feedback so your team can connect general customer sentiment with specific likes and dislikes.

There are also more tools than ever to capture *passive* feedback. Fullstory.com invented a way to measure rage clicks—a way to try to observe where customers clicked more than necessary to use a feature, which indicated they may have been dissatisfied. This sort of feedback is fairly blunt, but it could be a signal to your team to dig deeper into the potential issue.

Prompt for Direct Feedback When You Need It

Sometimes you need to be pushy and ask explicitly for feedback from your customers. This typically looks like a pop-up asking for feedback about the product.

Asking for feedback in this way is much more disruptive to the customer experience, so this method should be used judiciously. We advise a sampling approach to bother as few customers as possible. As long as you use a sampling approach, you can do it on a schedule, in specific key experiences, or when you launch new things. Reserve this method for evaluating a specific kind of feature or part of the customer journey that requires rich feedback that is highly valuable to your team, such as:

When you launch a new feature

Launching a new feature gives you two opportunities to solicit feedback. First, add a feedback icon on that feature that includes the opportunity for passive sentiment feedback in addition to full feedback. Then when your app observes a customer's specific use of the feature, activate a contextual pop-up (web app), notification (mobile app), or email (app agnostic) asking for feedback. Like all proactive solicitations for feedback, it should be done carefully and respectfully.

Churn experience

When customers are canceling their accounts (i.e., churning), you *must* ask for feedback. We feel strongly about this one. One of the strongest signals for improving your product is when customers make the effort to either cancel their account or downgrade to a free tier if it exists. It's important to fully capture the reason so your team has a chance to correct the issues driving churn. Remember, as long as customers maintain their account they can be reactivated with a particularly compelling message or feature, but account deletion is final—they exit out of your sphere of influence.

At WPEngine, Ezinne's team asks churning customers if they'd be open to speaking to someone in product about their experience and what led to their decision to leave. Churn feedback is now a very sophisticated innovation in SaaS and has become a fulcrum for reactivating customers. We will address churn in detail later in the chapter.

Onboarding experience

Onboarding experiences should be as personalized and *short* as possible. After onboarding is a good time to ask for explicit feedback to ensure customers feel prepared and that the process was a good use of time. The aim of this kind of feedback is to improve the onboarding experience for your product, which always needs refinement.

Often products will have a post-onboarding gamified checklist of key tasks to accomplish in the app. Customers rarely use these lists as intended, so it's still useful to ask what matters to them (note that A/B tests are highly useful here too). Onboarding is a critical first impression for new customers, so it's worth asking if you got it right and how to improve. For more on onboarding and activation, see chapter 4.

Post-activation

Activation is the stage when a customer has reached sustained, productive use of your product. Some customers will onboard but never activate, and many fall somewhere in the middle of activated and dormant. Regardless of customer stage, it's valuable to ask for feedback a few days or weeks after onboarding to gauge the customer experience. Post-activation feedback should be focused and personalized to the usage level of the customer. For example, fully activated customers should be asked different questions than dormant customers, and you shouldn't ask for feedback on a certain feature if a customer has not yet used it.

If you have a decently sized customer base, you need to have a sampling approach, especially if you're asking for feedback via email. Only about 10% of a customer base needs to be surveyed at any time. If you have hundreds of thousands of customers, then that sampling number can come down to 2–5%. Usually that's enough to get really valid feedback.

One exception to these sampling guidelines is when a customer churns. Companies should work to collect 100% churn feedback. It's so valuable that dropping any of it on the floor can be wasteful. If you're a high-performing software business, this number should never be higher than 1–2% per month, so the feedback shouldn't be too overwhelming.

Build a Community to Engage Early Adopters in Conversation and Facilitate Peer-to-Peer Support

For your early adopters, your product is usually essential. These customers also tend to be the most enthusiastic advocates of your product. Smart product teams harness this energy by creating a community of customers using groupware software.

This community, when carefully designed, can be a place for customers to support each other but also to have a conversation with your team about various aspects of your software. This conversation can be tuned quite easily toward feedback for your product team.

It's important to note a few things about communities.

The feedback you get from this community is early-adopter and advocate feedback. In other words, it's skewed. So while it is good to sense-check direction and other things about product decisions, it will likely not represent the mainstream of your ideal customer profile (ICP). As long as you understand this, the ability to get quick feedback should make up for this, in addition to listening to customers outside this community.

A community requires active management. Your early and most engaged customers will crave conversation and attention from your team. They become invested in your product and see themselves as co-creators. We highly recommend having a person on your team dedicated to managing the community. Hiring a full-time community manager can get expensive, especially for a startup, so we recommend hiring a part-time manager or recruiting community volunteers (even from the customer base) in the early stages.

Additionally, picking the right community tool is crucial. Companies can consider chat-centric options like Slack or more forum-focused software like discourse.org. We find the latter more scalable; but tastes may differ.

Get on Social Media and Respond to Your Customers

For some products, social media can function as a decent proxy for a community. There is often a segment of customers who use social media as a support, complaint, or hype channel. Remember that less than 10% of customers will ever say anything, so social media posts should be viewed as emotionally charged and treated accordingly: with care but not an overreaction. Like communities, social channels should be actively managed for two-way communications by a volunteer or dedicated resource at some point. Fortunately, there are many tools and automations that can help with this task.

Allow Your Most Enthusiastic Adopter to File Bugs (If You Can)

If your community is tech-savvy, let them file actual bugs that can be directly addressed by engineers. This is often the case for companies that do SaaS for developers and infrastructure engineers. It's important to strike a balance between ease of filing and friction so that trivial issues don't inundate the more important issues.

Set Up Bi-Directional Relationships with Customer-Facing Teams (Sales, Customer Support, and Customer Success)

Typically your customer-facing teams will focus on all kinds of potential and existing customers. This means that feedback the team gets should be highly prized. It represents the voice of your most valued customers.

Customer team feedback has a very high signal-to-noise ratio. At Calendly, the customer success and support team synthesized feedback every two weeks for the product team in a bi-directional meeting. They also condensed the most important product blockers for growth and linked them to individual cases and meetings so it was easy to see the frequency of these requests from customers.

Typically a CS team will have tools to track their customers, their meetings, and their feedback. The synthesis of these meetings is a gold mine of feedback, especially when taken together over a time period. If your team is lucky, they have a SaaS tool for this which may allow the data to be moved from there to Slack, Jira, and other triage and tracking tools for easy analysis.

BUILDING THE MACHINE

Product teams should have a multi-channel strategy for listening. Some channels are more important than others. For example, churn feedback should be weighted very highly, while sales signals need to be analyzed for context to make sure they are high value.

Different channels represent different kinds of stakeholders and customers, who provide feedback on disparate aspects of your overall business operations. *Only when taken together do they paint a strong pointillistic picture for a product team.* In fact, listening to only one or two channels inevitably leads to selection bias.

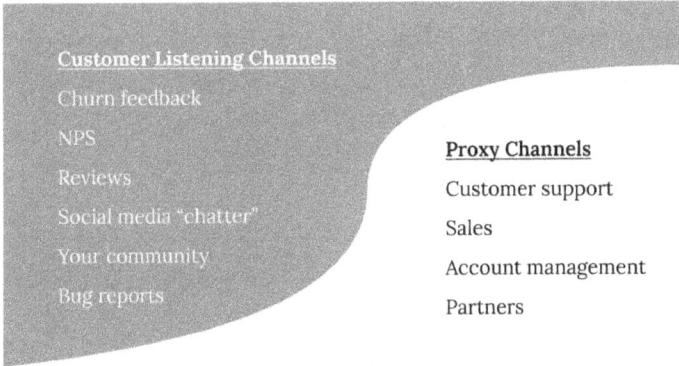

Customer Listening Channels
Churn feedback
NPS
Reviews
Social media "chatter"
Your community
Bug reports

Proxy Channels
Customer support
Sales
Account management
Partners

A good customer listening machine captures and synthesizes feedback from your most important channels and distributes the top insights throughout the company. Here is how we think about building such a machine.

1. Automate Feedback Capture and Triage

You can only grow so far without automating customer listening. Very quickly you have to try to automatically absorb it, organize it, and triage it. At Calendly we created a system to deal with the rapid influx of customer signals. It was not perfect and required jerry-rigging, but here is what we did:

1. Send all the data to Slack. You can use a tool like Zapier or similar and create separate channels for each feedback source. For example, #in-product-feedback channel, and #sales-feedback channel.

2. Also send data to a single Jira Project but with channel categories to enable filtering.

3. Give everyone access to all these channels. Allow them to add what they encounter in addition to the automated feed and empower the entire team to tag and annotate feedback.

4. Annotation can be used to promote high-signal content to a master feedback channel. In Slack you can build this automation via Reactji. I recommend that this promotion be paired with pre-assigning to a PM. Reactji can accomplish this easily.

5. This central feed can be also split into a different Jira project. Roughly a feedback triage project. Each ticket is assigned to the PM if tagged accurately.

This simple system will give you a river of feedback from all channels, annotated by the entire team for high signal.

The product team should own the process of annotating feedbacking and triaging. PMs should be assigned certain channels to review regularly. One important aspect of this triage is determining how often a certain issue arises. Issues that have the highest frequency and occur across multiple channels should be highly prioritized. PMs should then dispose of the issues in priority order: closed (not addressing), added to backlog (specific to a team), or attached to a feature (current effort).

2. Make Your Customer Feedback Searchable

We can never address everything we hear. However, when the team has prioritized a problem, it's useful to search the customer feedback archives to read the things they have already heard about this problem area. Doing so provides a rich sense of where it fits in the workflow of existing customers and potential customers. Additionally, if we have the context of the feedback, we can attempt to contact the customers to do some impromptu discovery if the elapsed time is not prohibitive to great feedback.

It's helpful if search is thematic (aka, fuzzy search) instead of exact match so that you can discover related concepts in the feedback quickly. Finally, it's helpful if search can show the frequency of a theme and how often it shows up across certain channels of a theme so that you can get an intuitive sense of how prevalent the issue is.

3. Publish Internal Digests

High signal-to-noise ratio channels should be delivered to the inbox of key leaders. We recommend a summary of thematic feedback across all channels should be treated the same way. The principle is simple: The more information that is shared about what customers are experiencing across the organization, the more essential the product and creative conversations can be and the more urgent solving the right ones can be as well.

At Calendly, Oji got a daily readout of the churn feedback from the company's cancellation survey. While Ezinne was at Bazaarvoice, the product team chose to create a Customer Voice council who were responsible for creating a monthly digest of their learnings across the different channels and sharing across the entire company. They did not have a listening machine built out, and this served as a stopgap while they built out something more programmatic. Team members (Design, UXR, and PM) rotated through the council each quarter, so everyone in the team had a tour of duty becoming intimately familiar with all the sources and tools that we could use to synthesize at scale.

4. Closing the Loop (When you Can)

Product-led companies should make every effort to close the loop by acting on—or at least acknowledging—feedback when customers bother to provide it. Customers appreciate these moments of reciprocation when they have expended effort and the company responds in kind. It humanizes the company and engenders loyalty. In short, it adds to product virality when this happens, and more importantly, it can lead to a better product.

Closing the loop on customer feedback is a bit like designing a product return system in ecommerce. For most companies it goes against the flow of all other operations. But product-led companies must become fluent in closing the loop; it is the only way to deliver with the speed and precision necessary to compete with the rocketships (and become one yourself).

Customer listening at scale is fundamentally about finding signal within noise. Not all customer feedback is valuable or actionable, but a lot of it is, and averaged together (using text and sentiment analysis), you can find real insight that represents an aggregated view of what your customers are thinking or patterns that require your attention.

Next, we'll discuss how product-led teams should think about turning their customer insights into well-designed products.

THREE

Product Design: Make Simplicity Your Biggest Weapon

Many product leaders believe design is a product-led company's biggest weapon, and in a way, it is. However, we believe what makes design effective is not font choices or brand colors—it's *simplicity*. Good design is detail-oriented, it's pleasing, and most importantly, *it makes products simple to use*.

Simplicity allows a customer to complete a desired workflow while making as few decisions as possible. It reduces cognitive load and makes software feel like magic, which in turn excites customers enough to tell their friends and colleagues. Our design philosophy can be summarized as: For the customer, "magical" lives at the intersection of *simple* and *powerful*.

Design is a serious investment that pays dividends in product-led companies and their products. The ideal ratio between designer and product managers in a team is 1:1. A significant investment in designers will help improve the threshold of design to be as good or better than the current level in the market. The cutting edge of design is always in flux (it's getting better all the time) and good designers will help your team be as close to the pragmatic edge without lagging behind. Trust us, customers will know if you are lagging; no matter how good your product, dated or clunky design will dampen their excitement for your offering.

However, mere design is not enough—design and product teams have to examine every pixel and interaction to strip it to its bare essentials needed to ensure good workflow outcomes for the customer. In the 1990s and 2000s—when technology companies sold software directly to the CIO and utility was the top priority—complicated customer experiences were the norm. Product teams added features, menu bars, and configurability options with abandon, hardly considering the impact it had on the end-customer experience (because they didn't have to worry about this). The outlier in the era was Apple, which demanded simplicity in its hardware and software from day one.

Microsoft, on the other hand, was the king of bloated user interfaces (UI). Old versions of Microsoft Office (which Oji worked on earlier in his career) had UIs that would make even the most seasoned back-office warrior faint.

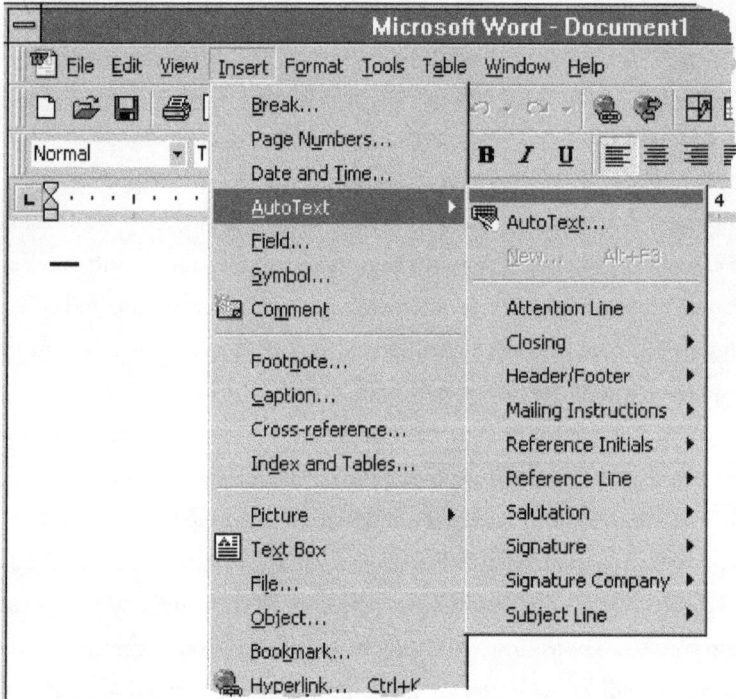

Trigger Warning: Old Office 97 menu system

Most of the options in Microsoft Office are used by fewer than 5% of its customers, but all were given equal weighting in the UI. The result was layer upon layer of menu bar options that made the products barely usable. Eventually Microsoft wised up and initiated a major overhaul of the Office menu system. Thus the ribbon was born, a new UI rolled out in 2007 as part of Microsoft's Fluent UI system. The main point of the ribbon was to visually prioritize the most common features and options within Office.

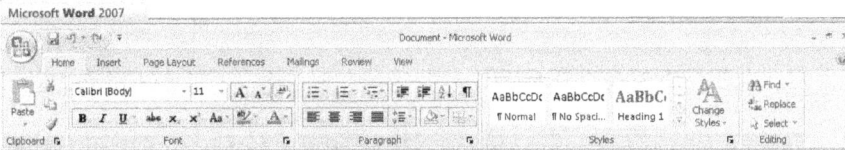

Office 2010 Ribbon system—more obvious

Eventually, Microsoft would apply the ribbon to many more customer experiences across Microsoft's applications. In fact, Oji was tasked with "ribbonizing" the entire suite of Microsoft consumer apps at one point—a frighteningly tedious but essential job that taught him plenty about customer experiences and simplicity.

Since then, many consumer and prosumer trends have forced software to be simpler and more approachable.

THE COST OF COMPLICATED DESIGN

Redmond, Washington, circa 2008: Oji was volunteering to teach older people at a shelter how to use Microsoft Office. It was a three- to four-weekend affair, teaching digital skills that would help folks find new and better jobs. Each day was a crash course in Word, Excel, Access, and Outlook.

For Oji, it was a very revelatory experience. Most of the folks were not familiar with computers and struggled with the most basic skills, like mouse use and keyboard literacy. But even the most adept struggled with the Office menu system and getting an extended task done. Many extended workflows required customers to navigate across different menus, and many menu names felt impenetrable. The cognitive load to get simple things done, especially in the harder tools like Excel and Access, was tremendous. For example, doing a mail merge in Outlook was next to impossible. Even as a Microsoft engineer, Oji couldn't do it in one sitting without some help. Microsoft had succeeded in getting Windows into a billion homes and workplaces, but actually using the software easily was another matter entirely!

This opened Oji's eyes to our assumptions as product managers and designers. Did we really understand our average customer? What else could we do to help them be more productive? It instilled in Oji the necessity of simplicity on product-led design.

THE ANATOMY OF SIMPLICITY

What does simplicity in design look like? Here is our definition:

Obvious, clear, and essential

Customers crave three main things in their user experiences (UX), in addition to good aesthetics: *obviousness*, *clarity*, and *essentialness*.

Complicated UX with massive menus are out. Obvious and clear UX that ease *cognitive load* are in. Clear UX speaks to design that favors elements (UI control placement, descriptive text, hover states, etc.), transitions, and function that is intuitive and minimizes customer confusion. If more than 80% of your customers of average intelligence can look at a user experience and intuit how to use it effectively, your team has likely scaled the bar.[9]

---

9   This can be tested in usability sessions with simple observation or tools like tree testing.

A NOTE ON TERMINOLOGY

User Interfaces (UI) are static pieces of graphical interfaces you can use a mouse of a finger (touch) to manipulate.

User Experiences (UX) are a sequence of UI-driven flows that help you go through a workflow.

A simple example: When onboarding into an app, the sign-up screen is a user interface. The question screen that asks about how you learned about the app is a user interface. However, the entire flow from sign up to question screen is a user experience.

Essentialness means the product has the simplest UI that can achieve the functionality desired. Distilling a product to its essential design often requires successive iterations. Many products fail the essentialness test by showing too many UI options too early in the customer journey. New customers get stuck because they simply don't have the information to make a good decision—yet. This is more acute in onboarding experiences. Usually, new customers are just trying to decide if the software is worth using for their workflow. Cognitive load makes them not activate and convert.

In the customer's exploratory phase, fewer options are actually better, and more involved options can be deferred to later stages of use, thus enforcing more simplicity. So you have to consider when controls should even be available. As an example, Coda.io is a powerful docs platform that lets teams create rich, multi-media workspaces and content. Its editor is intentionally bare and doesn't show any options until you write text. This focuses the customer on actually writing, trading static menus for contextual menus when you have actually committed some words to the page. Of course, you can turn them on if you want. (The Pro Edition of this book is built in Coda. See for yourself at productmind.co/brpro.)

REDUCING COGNITIVE LOAD

Minimizing the number of options for a new customer is a way of reducing cognitive load. Cognitive load refers to the amount of mental effort required to process information and complete tasks. In the context of product design, high cognitive load can lead to customer frustration, confusion, and abandonment, particularly for new customers who are still learning how to navigate and use the product.

To reduce cognitive load, designers should focus on presenting the most essential features and information first, and gradually introducing more advanced functionality as customers become more familiar with the product. This approach, known as progressive disclosure, helps customers build confidence and mastery over time without overwhelming them with too many choices up front.

Other strategies for reducing cognitive load include breaking complex tasks into smaller steps, using clear and concise language in UI, and providing visual cues and feedback to guide customers through the process. By minimizing cognitive load, designers can create more intuitive and customer-friendly experiences that encourage adoption and retention. Ultimately, the goal is to strike a balance between simplicity and power so the product can adapt to the customer's needs and abilities as they evolve over time.

Design teams should not be afraid to remove choices in their UIs that only a small percentage of customers are likely to make in order to deliver on simplicity. We address some of those trade-offs below in discussing highly configurable software.

The good news is that the emergence of web and mobile software has helped the cause for simplicity. Desktop software could be almost infinitely complicated. Desktop developers had a lot more layout control to map pixels to the screen in operating systems like Windows, MacOS, and Linux, which enabled this abundance of choice that rarely helped the customer. The discipline of interface design was young in the early days, leading to a lot of hard-to-use software.

In contrast, HTML markup (the primary language of early web design) offered fewer options and almost mandated spareness. It was almost too spare in the early days of the web, leading to a whole range of bridge technologies like Adobe Flash and Windows Presentation Foundation (WPF) that were used to make web pages more dynamic. Even as Javascript (the more standards-based dynamic elements of the web) became more capable, the simpler web ethos endured and began to pervade business software after conquering consumer software.

Obvious, clear, and *essential* are not necessarily *criteria* for a company's software's design aesthetic. Instead, they should be seen as *principles* for the ultimate outcome of user experience design. While minimalist design has dominated the last decade and encompasses these values, it does not mean you can't experiment with "maximalist" design trends, so long as the underlying user experience remains obvious, clear, and essential.

Remember, the tech industry has transformed from making software for experts to making software for everyone in this world. This requires us to consider ease of use so that all our customers can succeed.

Good Defaults

Complicated software UX and UIs stems from an inability or unwillingness to make decisions for customers (we can hear Steve Jobs yelling his admonishments from Valhalla!). Great product-led software simplifies the experience by choosing good defaults. The defaults should be the best choice for about 80% of customers. To determine what the defaults should be, a product team should use their customer discovery process to determine:

a.   What choices most customers will make
b.   What defaults will make most customers feel most productive
     the fastest

Good defaults can be universal or they can be adaptive, changing depending on the behavior or ongoing selections and product configurations of the customer. It can depend on previous choices customers have made or the information shared during onboarding. Often onboarding experiences can ask questions that help categorize customers on a scale of casual customer to expert. Questions to ask (usually multiple-choice) include:

1. What kind of customer are you?
2. What will you use this for?
3. What do you do most frequently during this workflow?

The answers can be used to set smart defaults that work for your specific customer. Some customers will definitely make different choices than the defaults. Usually, these are expert customers who are accustomed to fine-tuning the output and want to utilize the full power of the software. However, good defaults tend to satisfy the workflows of most of your customers without forcing undue complexity.

Opinionated UX

To build off the last point about good defaults, good UX is opinionated. Product teams should strive not just to augment existing workflows, but nudge customers into the most efficient way to accomplish the workflow. Opinionated UX can teach thousands of teams and millions of workers the *best way* to accomplish a certain task.

The opposite of opinionated UX is agnostic UX, which treats every product configuration as equal. Opinionated UX suggests a guided workflow that lowers cognitive load for most customers who may or may not know the very best way to get work done. However, the product should still allow expert practitioners to take control of the workflow if they need to and not stand in their way.

One way that modern business software does this is by allowing customers to be specific with what workflow they want to use. Atlassian's Jira can be configured to run either Scrum, Agile, or Kanban—three different opinionated workflows to manage engineering tasks. Customers simply have to choose their preferred workflow up front when setting up a project. Jira's UX is opinionated, but still gives customers choice on the type of workflow that works best for them.

All software sits on a spectrum from opinionated to agnostic. Product managers (and designers) in product-led software companies must choose where their product sits on that spectrum. In theory you can design infinitely configurable software that can accommodate any workflow and can technically be sold to any team on the planet. However, the cost is complexity, and only experts can effectively use your product. On the other hand, really opinionated software that doesn't solve for every workflow variation may be simpler to use but may not work for all target customers.

Our advice is to start with more opinionated products that guide customers toward the best way (given your customer sense) to accomplish work. This will force you to truly understand your customer's sharp problems and devise clever solutions for them. Once you find a successful opinionated workflow, add configurable power to your product so that a wider range of customers can customize their ideal workflows without sacrificing simplicity. Starting with agnostic UX and then making it more opinionated is much harder to pull off.

Powerful and Configurable UX

The holy grail for software is simple but also *powerful*. The goal is to balance your UI/UX so that it is clear, obvious, and essential, but *also* gives room for expert practitioners to precisely tune the workflow and push the tool to deliver on demanding task parameters. Most of your customers will not be expert practitioners, according to the Pareto principle, so temper this advice based on the specifics of your customer base. Remember that expert customers also want good defaults that lower cognitive load. They just want more options and control when they need it. We generally recommend that companies should start with *simple* UX and evolve toward more *expert and enterprise-level* power and configurability.

For example, Calendly organized its settings so that 80% of what customers needed was in a single window called "event details." But they also offer a suite of advanced features for configuring the meeting and scheduling options exactly to the customer's exacting preference. Calendly hides those advanced features in separate pop-up windows, available if power customers need them, but not distracting for everyday customers.

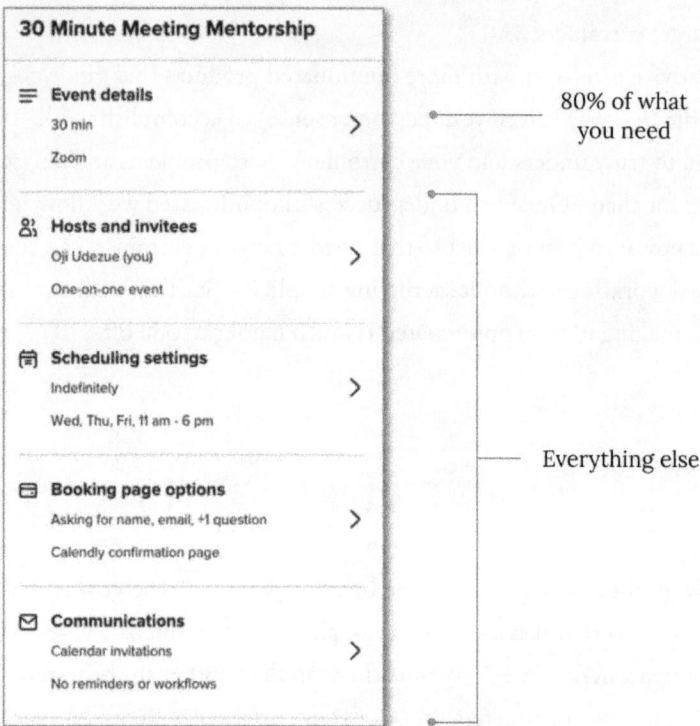

Another example—in an entirely different industry—is Tesla. Most Teslas are sold with a standard package: 2-wheel or 4-wheel drive, basic software package, etc. But for Tesla's richest or most-demanding customers, they can upgrade their vehicles to Ludicrous Mode, which sacrifices some of Tesla's famous simplicity for world-beating performance.

Whether it's software or cars (which, today, are essentially one and the same), simple is the best default, with advanced settings readily available for expert customers. Once a customer bypasses the default settings, they become responsible for their own results.

SIMPLICITY FOR STARTUPS: SHOULD YOU BUILD AN MVP?

People often talk about building minimum viable products (MVPs), but the concept is often misunderstood and can lead early product teams astray. Part of the difficulty with the idea of an MVP is the definition itself: What is the *minimally viable* version of your product? This is an extremely difficult question to answer when you haven't dialed in your target customer definition and don't even have a product yet.

The first problem is clearly defining a minimum during an early stage. The other problem is about viability. Can you create an MVP people enjoy using? Many startups get stuck with a bare MVP that does not let them learn and which no one can use happily. Most of your customers won't get hooked on a bare-bones product that hardly works unless it's groundbreaking.

There is a lot of value in building an MVP *for learning purposes*—feeling out the problem and how sharp it is. For example, when Oji was building Intermingl—a startup for connecting people around shared interests and live events—he built a simple software MVP to prove the value of the workflow. He used it in early customer surveys and to pitch investors on the vision. There was no assumption that version would be the backbone of adoption. Once that was done, he discarded that version of the tool entirely and built something stronger from the ashes. A more famous example of a good MVP was Drew Houston's first "version" of Dropbox. "Version" is in quotations because there was no working software; Drew mocked up the concept in Adobe and created a video to show off the workflow.

For building an *actual* first version of your product that you plan for your customers to adopt, we prefer the acronym SLC: simple, lovable, complete.[10] WP Engine's founder, Jason Cohen, created this concept and it has gained traction because it's more conducive to product-led thinking. Building an SLC requires focusing on a single core workflow and ensuring the product solves a real problem in a simple and loveable way. Once you attract early customers with your SLC, you can begin to build out the rest of your product vision.

So bottom line: MVPs are built for learning and SLCs are built to ship. The two may not even come from the same code base. Be careful not to invest too much time in an SLC before collecting enough insights from your target customers via your MVP.

How to Keep It Simple: Appoint Champions for Simplicity

Every product needs *champions of simplicity*. People who can reject the first iteration of design and demand that more simplicity iteration be done on the user experience. Usually, champions of simplicity are either product managers or designers. The job of the champion is to model and enforce simplicity until it becomes part of the product team's DNA.

The champion of simplicity need a modicum of authority—a way to veto whether user experiences meet the appropriate simplicity bar. This could be done with a mandatory design review or something similar. Adding this review process is also a way to embed the principle of *simple* into your company culture, which is the ultimate goal.

Simplicity must be demanded from the moment a customer comes in contact with a product. In the next chapter, we will look at how to create a simple and effective customer activation workflow.

---

10   Cohen, Jason. "Your Customers Hate MVPs. Make a SLC Instead." A
     Smart Bear. https://longform.asmartbear.com/slc/.

FOUR

Onboarding and Activation: Getting to "Aha"

In-product customer onboarding and activation is a relatively new art in software businesses. Traditionally, we either did not do it at all (expecting customers to RTFM[11]) or used sales and account management employees to accomplish it. However, as the cost basis of building valuable software has come down and alternative direct-to-customer go-to-market approaches have proliferated, technology companies have sought to scale the introduction phase of their products via software, reducing customer activation costs.

Onboarding and activation is the process of crafting the best possible first-run experience for a new customer so that they see your product's value quickly and deem it worthy to pay for. Successful customer onboarding and activation result in regular, productive, and paying customers. Poor activations lead to low adoption rates and increasingly high marketing spend.

Before the rise of cloud computing and PLG, most B2B products were sold to CIOs—they were the proxy for the actual customers inside the company—and thus the task of software selection for workers fell squarely on the IT department's procurement team. The CIO's team's tastes overruled those of the people who would use a product.

---

11   Read the effing manual.

Once acquired, the IT department would turn around and roll it out. Many IT departments struggled to get all their users activated and productive without the enterprise investing in some expensive software training—think of this as an onboarding or activation budget.

This "activation gap" even became a profit center for many enterprise software companies, who, for a few dollars more, would deploy some consulting services to help train the end-user customer for the CIO. Many software companies made a *lot* of money from these consulting services (and some still do).

A lot has changed and is changing from that world. As business consumers have started to assert their purchasing power separate from the CIO, business software has had to become more intuitive and usable, since its utility is tied to the satisfaction of the end user and the augmentation of their primary workflow.

Additionally, *professional services* are falling out of favor in a competitive software market. Many companies now have to show customers how to use their products for free, because as the purchasing behavior shifts, the end users don't really have an onboarding budget. This makes it more costly to sell software that is hard to use immediately, and why simplicity is one of a product's most important features.

The new default purchase workflow often looks something like this in the low-to-mid end of the market:

1. A business customer hears about a new software tool that colleagues or other professionals in the field are using (often via social media or in-person word of mouth).

2. The customer visits the company's website to see if the value propositions align with what they're looking for.

3. If so, they sign up for a free trial (or they may follow the brand on social media and plan to return later when they have more time to review).

4. Once signed up, the customer goes through the onboarding process.

5. They upload a preliminary data set to test out the product and see if it's worth their time.

6. If the results are compelling, the customer will use it to do their work and test the output on their coworkers and teams.

7. If that works out well, they keep using it and either:

   A) buy it directly, if they have buying authority, or B) ask their boss to pay for it.

8. If it's a multiplayer tool, they may invite the rest of their team to use it.

Note the total lack of direct sales to the CIO and professional services in this model.

Some technology sectors still rely on the CIO-mediated process. Slower-moving industries like healthcare and insurance are two examples. This is an opportunity for ambitious startups and product people; just because product-led is "the new way" does not mean every industry has already adopted it. Just like every technology trend, it is not evenly distributed across the economy.

Let's explore some of the finer points of this product-led process of acquiring new customers and bypassing the traditional product acquisition gatekeepers.

PRODUCT-LED ONBOARDING, ACTIVATION, AND CONVERSION

Many companies use the terms onboarding, activation, and conversion interchangeably, but they are all distinct stages in the process of turning a prospect into a customer. Here's how we define each stage:

1. Product onboarding is the process of helping customers see the value of your product by experiencing the target workflow. The goal is to help them reach the "aha" moment.

2. A customer is considered activated when they reach sustained, productive use of your product.

3. Customers who have paid for your product are considered converted. Customers usually convert post-activation. But it's possible to pay before activation[12]. Post-activation customers have decided that you are their tool of choice for the problem they aim to solve for themselves.

4. At any point the customer may come to the conclusion that the product and its features are not fit for their purposes, and they abandon it. These customers are considered to be inactive.

5. Converted customers who stop paying for your product are considered to have churned. (Conversion is the subject of chapter 5.)

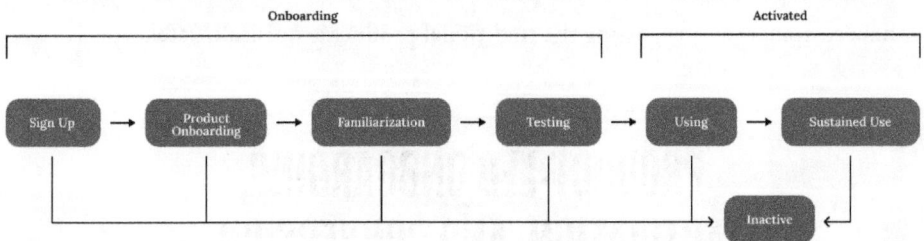

Onboarding is becoming completely customer-directed. In fact, the need for any in-person chat (or, God forbid, a customer support call) is considered a failure

---

12 Very common during the CIO buying era.

and often leads to lost customers. Poor onboarding leads to failed activations, and failed activations lead to lack of conversations.

In our opinion, you cannot spend too much time optimizing your onboarding experience. It should be lean, clean, and effective—any wasted effort could result in failure. Let's dig in to what makes good onboarding.

THREE INGREDIENTS FOR GOOD ONBOARDING

Onboarding software prepares customers to use your software for the first time, with the goal of activating them into a sustained, productive user. It's an entire workflow in itself, designed to set new customers up for success before they access the main application UI/UX.

The mantra for product-led software in onboarding is time-to-value. The goal is to get to the moment where the value of the product *clicks*—often referred to as the "aha" moment—as fast as possible.

Keep It to the Essentials

The guidelines for good product design also apply to onboarding—simplicity. Hence, the first ingredient to good onboarding is *essentialism*. Most new products require the customer to make some *critical choices* in order to work properly once you start using it, otherwise the first experience will be confusing and useless. Product managers and designers need to be critical about what is truly essential during onboarding and what is not.

Even the necessity of the sign-up process itself should be questioned. Some product-led companies are skipping sign-up entirely so customers can begin to experience the value of the product immediately[13]. For example, most project management tools could be used without forcing the customer through a sign-up process, as long as it's intuitive to start. Only when they want to save their work or share it with teammates would they need to create an account—this is after the customer is invested. We will continue to see more products move in this direction as the essentialness of each step is heavily scrutinized.

Of course, the sign-up process *is* essential in many cases—for example, at Calendly the sign-up phase was how the team established authentication to a customer's calendar (usually on Google or Microsoft services). Calendly requires just a few choices before a new customer can begin to use it: connecting to their calendar, creating their Calendly link, and setting their default availability. Not only are these necessary choices to be made up front, but the customer needs to be oriented on the meaning of these concepts to use the software properly.

Some teams also need to filter out customers who are likely not going to activate, like those trying to use a business product for personal use. Forcing sign-up with a corporate email and actively gating personal email sign-ups may be a necessary step, but it should not go unquestioned.

Every Step Should Make Sense

This leads us to the second ingredient of good onboarding: cognitive orientation. The steps that are deemed essential should be *wrapped in context* so the customer understands why they need to make that decision before they begin to use the product.

Cognitive orientation means providing a set of lightweight explanations of key ideas and concepts that will make activation more likely. This can come as part of the essential pieces of onboarding but also through the rest of the user experience.

---

13 Not forever, just until the customer is deeper in the actual relevant
activation flow.

Cognitive orientation is a different kind of required thing—it's framing that will make the first week of using the product more intuitive. In the case of Calendly, we introduce the idea of a meeting link (a digital space where people can go to figure out how to schedule time with you), the idea of how much time you want to make available to schedule in the first place (availability), and how Calendly works with your cloud calendars to support your scheduling workflow. It helps make clear that Calendly isn't actually a calendar application; it does *scheduling* for customers' existing calendars.

Cognitive orientation can be active or passive. Active orientation can look like a virtual product tour that guides the new customer through key functionality in the product. Some onboarding workflows *gamify* onboarding by displaying a list of key tasks to complete at the top of the application. Customers are reminded of these tasks each time they log in. Passive orientation includes adding info bubbles to certain features, where a customer can click on them to learn more within the app.

There is no golden rule on how to provide cognitive orientation, only that it should be pared down to the most critical points (see the first ingredient). Anything else becomes noise that can distract the new customer from completing onboarding and activating. Some products require a longer setup, and in these cases the product team should set the proper expectations with the customer up front. For example, if linking social media accounts is core to the product workflow, be sure to include a disclaimer early in your onboarding.

Onboarding Is a Conversation

The third ingredient of good onboarding is an extension of cognitive orientation: a product's *messaging strategy* during onboarding. Customers should be reminded in-app and via email and text message to complete critical onboarding tasks that lead to the aha moment. Your messaging should feel like an extended and seamless part of onboarding UI.

Most messaging strategies are too simple to be effective—they send out a set schedule of emails on day 1, day 3, day 5, etc. We recommend something more sophisticated—an onboarding messaging campaign that is responsive to, and conditional of, customer actions and engagement. This kind of system can then be eventually connected to other non-onboarding experiences like reminding customers to pick up where they left off, or notifying them of urgent team requests via texts, etc.

Make a Distinction Between Required and Helpful

In general, product teams should draw a firm line between required things and helpful-but-not-essential things in onboarding on the way to activating a customer. The required things should be as few as possible and over time, even fewer through active measurement and pairing (except when you missed a critical element). We recommend taking your first list of required steps and cutting it in half. Be extremely brutal. Required things can be legitimately put in the path of all new customers in an un-dismissible onboarding experience.

Helpful-but-not-essential things should be provided with a nudge approach. These tips should not get in the way of the customer (i.e., they are dismissible and not demanding of all their attention), but still catch their attention and encourage customers to complete them. Over a few first sessions, a well-designed nudge UI can be very effective.

Essential	Helpful
Sign up / sign in (in most cases)	Marketing survey e.g., "How did you hear about us?"
Customer details, company, role	Team invitation
Context on each step	Step-by-step UI walkthrough
Critical initial choices (e.g., default hours)	Selection of the right template

Example lists of essential vs. helpful onboarding steps.
Your actual lists may vary.

INVITING TEAMMATES AND COLLABORATORS

Most business software does not work well for a solo employee. They are often team tools. Think about Jira or Slack, for example—you need other people using the tool to gain value from it. Usually there are three or more kinds of people using the same software as a team: administrators, collaborators, and viewers. However, in almost all cases, the first evaluator has to feel confident enough to invite others to join their evaluation of the team experience of the software. So while invitation is a long-running idea within team-focused business software, it takes on special significance during onboarding. Customers are unlikely to activate onto team-based software unless they test it out with their teammates first.

(Note: A strong single-player mode can sidestep some of this dilemma for team software, but that is not always possible.)

Using the principles we have talked about above, here are some key ideas on invitation as part of onboarding.

Invitation Is Not a Required Thing, Unless it Really Is

We tend to slot invitation into the helpful things category. The reason is that good invitation UI can be quite involved.

When individuals start to evaluate your software, it's best they come to a conclusion about its value first before bringing on a team. Humans are funny, we don't like to look bad in front of our colleagues, and we want to be confident in what we recommend. Prompting customers to invite other users too early in their exploration of the software can be off-putting. Some teams split the difference and construct an activation flow that invites at least one other customer very early on. This is distinct from the general invitation UI.

If inviting another user is key to getting to that aha moment, then the invitation should be part of the main onboarding flow.

Invitation Should Be Lightweight and Intuitive

Invitation UI should be lightweight and—like all great UI—*simple*. It should be easy to invite others.

- Allow comma separated email addresses.
- Allow selection of names from the internal directory (make it possible to directly authenticate to the local directory, so there is no need to hunt for emails to invite collaborators).
- Show people who have been invited and their invitation state, including the ability to re-invite.

The invitation UI should be available on demand—at any time a customer should be easily able to invite one or more teammates to share in the app or in a project in a consistent way. Product teams can also use their messaging strategy to nudge customers toward inviting teammates at different times during and after activation.

PERSONALIZATION

Even the most well-designed software can benefit from personalization—allowing power users to tailor the experience in a way that feels most useful for them. Personalization is not all that common or easy to do in business software yet. However, a couple of things tell us that this is going to change quickly:

1. The scaled success of personalized consumer software (think TikTok's For You page)

2. The arrival of generative artificial intelligence with new personalization capabilities

We can only expect the rise of personalization to continue for business customers.

Personalization in consumer software is usually tied to the customer's preferences (for example, the type of videos they like to watch). Personalization of business software, though, is usually tied to the customer's identity and usage patterns. This is why many product-led products ask about a customer's job title and role during onboarding. This information can be the beginning of a personalized experience. Product teams can also learn about the customer based on their everyday use of the product.

Note that we don't recommend asking too many data-gathering questions during initial onboarding—again, the first rule of product-led onboarding is essentialism. Strive to make it as short as possible in order to speed the path to activation. In addition, market data questions are indulgent; they help your team but barely help the customer (at least in the short term).

Finally, note that hyper-personalized product features can offer diminishing returns (see section below for more details) if it's not harmonized with the need for simplicity.

Personalized onboarding can be applied to many things, but following are a few personalized experiences we recommend.

Default vs. Advanced Features

Giving expert users the option to configure advanced features during non-essential onboarding steps will help you satisfy your most demanding customers from the beginning. Use info bubbles or product hints to show customers where to find advanced settings. Most customers will simply ignore or close the hints, but your power users will eagerly "pop the hood" to see what your product can do.

The introduction of advanced features can continue after onboarding as well, so don't feel the need to introduce everything all at once. Recommendations can be made based on usage patterns after the customer has had time in the product. This is an excellent way to balance simplicity with more advanced features in an automatic way for customers (less effort). Refer back to chapter 3 as you balance simplicity vs. configurability.

Preferred Workflow

Some products offer multiple "default" workflow options. A Jira project, for example, can be configured for Scrum, Kanban, bug tracking, or a basic to-do list. Introduce the various options to customers during onboarding with a bit of education to help the uninitiated make the correct choice. A pop-up product carousel is a great tool to briefly explain each workflow option. But even more powerful is when a product will understand the customer's preferred workflow and automatically default to it unless asked to show the full complement of workflows (for exceptions).

Customer Support and Assistance

A key element of personalization is customer assistance. What kind of help does this kind of customer need when they are on this UI page? Customer assistance used to be blunt, like a help button on every page. But it's important to know where certain kinds of customers tend to get stuck and provide contextual support right in the application, in the form of suggested knowledge base articles or videos.

Knowing when, where, and what type of support to provide goes back to building a customer listening machine. You want to know what obstacles new customers run into and place helpful guides or tools along their path.

When NOT to Personalize

But personalization is not an undiluted panacea for great product experiences. As we discussed in chapter 2, hyper-personalization comes at the cost of simplicity. The key is in the tailoring. How much personalization is needed? At what cost?

To what benefit? Personalization is not usually an early product cycle endeavor anyway—usually there is more important low-hanging fruit.

We recommend adding personalization once your product has reached the optimization and growth phase, well after it has achieved product-market fit.

Here are some reasons why:

- **Personalization is resource intensive.** Building personalization is time consuming and expensive, which is why most early products should lean toward being opinionated instead of configurable. There are new tools on the market that help software build personalized experiences, such as Stonly and Userpilot, but they are still relatively immature. (But keep your eye out on these tools; they are rapidly becoming more powerful and easier to use.)

- **Experimentation matrix explodes.** Applying personalization to product growth initiative makes focusing on statistically significant product experimentation harder. Each personalized product workflow needs testing, and if there are too few customers using that path, it can be hard to experiment on it individually.

  Because changes often need to be applied to every customer segment, the experimentation matrix can become quite large. How much did this change have on this segment vs. that? And if the impact is uneven in different segments, should it be deployed? These are not easy questions to answer, and that is just the tip of the iceberg in this case.

- **You need a critical mass of customers for it to matter.** Early products, in their quest for product-market fit, tend to hyper-focus on a single persona. With that comes modest goals like acquiring the first ten thousand customers. Therefore, personalization is not the highest need and may hurt the company's chances of finding PMF. The only exception is if personalization itself is a key feature of the product rather than an aid.

The siren song of personalization is hard to resist for early product teams. Customers asking for personalized features are often loud and insistent. It would be easy to build what they want. But as they say, building a product for everyone leads to a product for no one.

CHOOSING THE RIGHT CUSTOMER ID TO MAKE GROWTH FASTER

One of the most dangerously overlooked aspects of activating customers is choosing the right customer identifier. We've seen this simple decision cause so many headaches that it is worth having its own section here.

Steve Blank (well-known investor and our former instructor and professor at UC Berkeley) famously said that a startup is an organization searching for a repeatable business model. With an emphasis on "searching"—meaning that startups should optimize for opinionated flexibility. In service of this ethos, a lot of the scale advice for startups in the last five years has been admonishment to do some things manually and defer scale until they find that optimum product-market fit—you nail it and then you scale it.

The downside is that you may make decisions early on that make scaling extremely difficult. We know many fast-growing companies who made bad architectural decisions that haunted them for years. And for some, those foundational mistakes took them off the chess board.

One such problem we have seen happen in recurring fashion, across multiple startups now, is the choice of a primary identifying key for your customers. A lot of companies use email as the primary identifier for customers. Thus customers use email to log in and identify themselves. In many cases, they can only use one email identifier at a time, which serves as their database ID. Examples include Atlassian, Amazon, and many, many other business tools you use and love.

Don't do it. We don't care how common it is or how many startup bootstrap frameworks work like this. Using email as an identifier will hamstring you and cause massive headaches down the line. We have seen multiple massive, world-renowned companies spend tens of millions of dollars just to re-engineer their identifier architecture. Eventually, every company needs more flexibility in how they organize customers and the identity framework underpins everything.

In B2B SaaS specifically, a lot of customers use their company email as an ID. As they move to different companies, they take good tools with them. But because company email is the ID, they have to ditch their old account and sign up with a new email at the next job. As a result, the customer loses their history and personalizations of the software. The product company also loses a way to connect and communicate with the customer and influence their retention during this delicate interregnum where they could always choose a competitor.

The other problematic pattern we have seen is that in general, people have multiple email addresses and numbers. Some of these IDs are backed with real data that can and should be connected to your software to deliver a better service. Accepting only one of these as an ID can be suboptimal for a software company and its products.

You should design customer ID systems to maximize these and other opportunities for your software and your product funnel. Here's our suggestion:

The best way to construct an ID system for your customers is by using a unique unguessable random number[14]. We generally call these globally unique IDs or GUIDs. Then you can attach all kinds of other identifiers to a customer's GUID to create a flexible system of identification. Multiple emails, phone numbers, and even other GUIDs can be associated with *and* dissociated from the primary GUID. Your customer doesn't even need to know their GUID or interact with it in any way—they can just use any of the associated IDs to log in and configure the software to their liking.

---

14 Unguessable by machines/computers, not just humans.

If your customer joins a different company, they can simply add their new email and jettison the old one. It makes no difference to your identifier system because their GUID does not change. Your customer doesn't need to provide an email at all—they can use a username, phone number, PIN, or any other identifier. It's better for the customer and better for your company.

Here is what it could look like (among other possible schemes):

GUID (primary identifier)	
Email #1	can login
Phone #1	can login
Email #2	cannot login
Phone #2	cannot login
GUID #2	cannot login
Name (Fname, Lname)	can login
Address	cannot login

In this example, you have two emails connected to an account, as well as a phone, simple username, etc. One email[15] can be used to log in, as well as the phone number. The benefit of this setup is this: If the product uses calendar or email data (or can send email on behalf of the customer), your system can use the

---

15   As a matter of good security, using multiple *email addresses* to authenticate and authorize may be a bad idea because one may be compromised when you're not using it regularly. Best to use one of each to authorize at a time. And the customer can decide which in their security settings.

data represented by any of the email IDs to deliver on the customer promise—even though the customer uses only one of those to log in. For example, I can send email from multiple Gmail addresses within the same account, even though I can only log in with one of the emails.

This is not rocket science, but I still see so many companies make this error. Don't be one of them.

Customer onboarding and activation is a game of trade-offs. It's a balance of personalization and opinionated UX; of simplicity and power; time-to-value and maximum value. It's a never-ending process that the best product teams return to again and again as they learn more about their target customers and how they use the product.

Once a customer is activated, the fight for their dollar has just begun. Next is how product-led companies think about converting customers into paying fans.

FIVE

Converting Customers: From "Aha" to Cha-Ching

In a software company, fortunes are made by getting thousands and millions of customers to pay for your product. This crucial step is called *conversion*.

For most of the history of software, *marketing* and *sales* were the key to customer conversion. They launched new business, educated prospects & nurtured relationships, closed deals, and collected checks, preferably from deep-pocketed large companies. However, product-led companies have engineered an additional new way to convert customers: a low-touch (or no-touch) process that happens completely through product-based automation. Prospects sign up for product trials, onboard themselves, activate themselves, and set up a payment plan—no human interactions necessary.

This is the holy grail of profit making in all kinds of companies, especially technology companies, because companies that can rely on this second version for profitable growth can lower their customer acquisition costs. In-product conversion is also very efficient; it's always working, even when your salespeople are sleeping. It even works in international markets automatically, especially if the product is able to be used in local languages and is supported by local payment methods.

As sales and marketing expenses fall, profit margins soar. Add in the near-infinite margin scaling of software in a large market and things get even more exciting. The economics and velocity of customer acquisition in the product-led growth model is why there are so many *more* unicorn startups today.

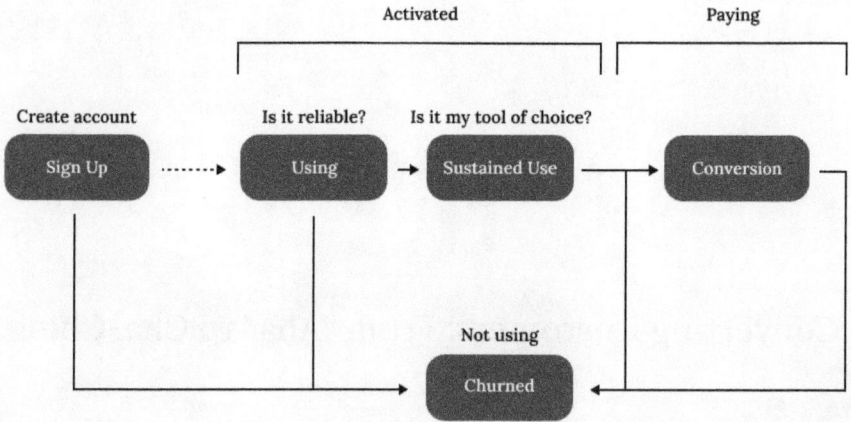

It's important to note that good product-led companies don't eschew sales entirely. We've mentioned that our definition of product-led growth is about matching customer value and buying patterns more exactly than past generations of technology companies.

Product-led companies don't see marketing and sales as a blunt instrument like in past cycles. They optimize conversion in the product itself and reserve sales for where it is most effective. They align sales and product-led methods so they work hand in hand, not in opposition to each other. New prospects can sign up for a free trial, activate themselves, and even invite their teammates without ever talking with a salesperson. However, a salesperson can inspect the account and find they are a really gigantic company and may need more help to bring their entire department or company on board, and engage to do that . . . and win the big sale.

We refer to product-led to sale-types of prospects as *product-qualified leads* (Pqls) and they are much higher quality than your traditional sales-qualified (Sqls) or marketing-qualified leads (Mqls). Customers have already started voting with their feet (and sometimes wallet) by *using* the product—a very valuable and important signal that is missing in the other types of lead-gathering methods.

PRODUCT-LED CONVERSION

So how do you get customers to pay for your software without a human involved in most cases? In a nutshell, you activate them effectively (see the previous chapter) and then *nudge* them into converting to a paid plan or premium features. Doing this at scale requires *generosity*, a sense of *psychological safety*, and *low friction*:

- Great products are **generous** in giving new customers access to features they have not paid for (yet) in order to help them build conviction in buying the product.

- The customer must feel **psychologically safe** to evaluate the software without the risk of being charged or getting locked in.

- It's the responsibility of the product to **remove as much friction as possible** in this process. Any confusion or lack of clarity could make the customer think twice and exit the buying journey.

Product managers should think carefully about features—which should remain free (if any), and which should be paid—and how and when these features are exposed to the customer in order to convince them to pay/subscribe. This orchestration starts from the very beginning of the customer journey.

We will first examine various methods to introduce product features to customers in the onboarding process. Then we'll look into how these methods can nudge customers to convert.

Product Trials

One of the objectives in PLG product design is to get the customer to try your product as quickly as possible, as seamlessly as possible, to help them understand if this solution is the one that works for them. We have even advised you to consider delivering value before there is an account sign-up. Software trials are ubiquitous today and generally considered a best practice, but there are a few different approaches.

Sales Trials

In the earliest days of software, every trial was a sales trial—also known as a dead-end trial. Despite building cutting-edge technology, virtually every software company was sales-led, meaning they built out large sales organizations to generate leads and close deals. The process usually involved multiple phone calls or in-person meetings, where salespeople would qualify the prospect and determine their needs. Then a custom software trial would be built and the salesperson would walk the prospect through it. Then the prospect decided if they would buy the software or not.

The sales trial model made sense in the early days of software because education was a massive part of the sale process due to low market penetration of software. Today, though, sales trials are the antithesis of product-led growth. Even in heavy sales-led models, an easy path to experience the product is very beneficial.

Traditional sales trials are the least generous, least psychologically safe, and highest-friction trials available. They offer very little in return for a customer's precious time, energy, and most importantly, trust. Customers crave autonomy in the buying process, and sales trials force them to operate on the vendor's terms, not their own.

Sales trials are clearly out of date, but many large, slow-moving enterprises still operate this way.

Time-Limited Trials

In the 1990s and 2000s, as software became more ubiquitous and consumers more knowledgeable, software companies developed the timed trial, known better as the free trial. In this trial, new customers get to use the product for free for a certain period of time. After the trial period, customers will *lose access to the product* and will be prompted to purchase the software. Customers could try calling the company to get an extended free trial.

Time-limited trials are still popular in software today and can work under certain conditions. They are most effective when targeting a narrow, high-value niche customer. A recent example is a product called ChatPRD, an AI copilot for developing product requirements documents. Claire Vo, the founder of ChatPRD and three-time Chief Product Officer, offers three subscription tiers and a free trial—but no "free plan" (as we'll discuss in the next section). This model works well for a product like ChatPRD because it's not likely to attract casual customers who need more time to evaluate.

The length of a free trial period has to be chosen carefully. It has to be enough time for most customers to get to the aha moment. Usually, you can interpolate this by looking at usage data and constructing a distribution of time-to-subscription for your product. In practice, most product teams opt for some multiple of weeks: 14 day trial, 30 day trial, etc. Usually trial periods are accompanied by automated email nurture campaigns that urge using the software and create a sense of FOMO during the trial period.

Free trials are more generous and create less friction than sales trials, but they are not entirely psychologically safe because they put the customer on a clock with the implicit threat of loss of access. While this can create a sense of urgency, the opposite is also possible, where many customers grow frustrated or irritated and try a competitor.

For products with a wide top-of-funnel of evaluators or in a competitive segment, free-trial approaches can be hit or miss. Free trials do not correctly value customers who do not pay you immediately. What is the value of a customer who converts on day 29 of their trial vs. one who converts on day 364 of their trial? If the LTV is high enough and the cost of serving a free customer is low enough (and it's essentially zero for software products), there is still upside in serving customers who don't pay for your product immediately, but could potentially become converts at some point in the future. That brings us to the latest iteration of trials, made famous by product-led growth companies: freemium.

Freemium (and Reverse Freemium)

Since the late 2000s and early 2010s, we have lived in the "era of convenience" for software. Ease of use and customer experience are prioritized over all. This customer-first mentality has popularized the freemium trial.

A freemium trial is one in which customers can evaluate your software in perpetuity at no cost. Companies create a free tier option for their software so that new evaluators can use it and have endless opportunities to activate and convert. Thanks to falling costs across sales, marketing, engineering, and backend infrastructure, the economics of the freemium model are more enticing than ever.

Freemium works for several reasons:

1. Customers can easily create an account and onboard themselves—activation.

2. They have unlimited time to evaluate and hit the aha moment.

3. They can still be nudged to convert fairly aggressively by your product, sales, and marketing.

4. The value represented in the free and paid tiers of your product can be varied, via experimentation, to achieve optimal activation and conversion characteristics.

5. As the product becomes more capable, it's possible to adapt your pitch to your free customers to convert them, as more enticements (enhanced features) become available. This is as opposed to re-acquiring them with a much higher marketing expense.

A version of freemium, commonly referred to as **reverse freemium**, is even more compelling. In this version, customers are *onboarded to the most expensive tier of the product* for a trial period. All the functionality of the product is available to the customer for a limited time. When that trial period ends, the customer is downgraded to the free tier, meaning they still have some access to the product but lose the premium features. The calculation is that some higher fraction of customers in the trial fall in love with your premium features while experiencing the full power of your product. When the free trial ends, they are more likely to feel a real sense of loss, which can quickly translate into a conversion.

Reminder: Simplicity is key to making reverse freemium work. If your most advanced features are not simple to use, it will be impossible for new evaluators to explore them and get hooked.

THE VALUE OF FREE-TIER CUSTOMERS

The freemium trial model allows your potential future paying customers to keep using your product until they decide to pay you. Free customers can be viewed as pre-primed leads with high revenue potential, and they should be treated as such.

According to a dissertation paper by Clarence Lee, Vineet Kumar, and Sunil Gupta from Harvard Business School[16], there are four specific benefits for having a generous free tier:

1. Continued usage: Customers who use your product in personal mode are the most likely to become paying customers, especially if the free/paid divide AND subscription points are designed very carefully using data and human psychology.

2. Virality: Satisfied free-tier customers can contribute significantly to your product's word of mouth, almost at the same rate as paid

---

16 Lee, Clarence, Vineet Kumar, and Sunil Gupta. "Designing Freemium: A Model of Consumer Usage, Upgrade, and Referral Dynamics" (Dissertation, Harvard Business School, 2013), https://www.hbs.edu/faculty/Pages/item.aspx?num=45458.

customers. If you have a product good enough to generate virality, this is worth a lot of free referral and consideration.

3. Social and team sharing: If your product can be shared socially or used in a team even while free, free-tier customers can introduce it organically to other people who will consider it for their tool of choice and could become paying customers.

4. Investor proof points: Free-tier customers often provide a proof point for investors early in a startup's lifecycle.

In short, your free-tier customers have significant value to your growth as a company, especially if they create virality or refer other customers. Oji certainly found that the LTV of free customers was positive at Calendly.

DESIGNING THE REVERSE FREEMIUM TRIAL

While not outright disqualifying the other two options, the freemium approach—and specifically the reverse freemium—is the best default for many mid-market customers.. It provides the highest levels of generosity and psychological safety and by far the lowest levels of friction. The customer is completely in control, and instead of looking like a greedy corporation, you are giving away real value for free. This goodwill could pay dividends over time.

Let's get into the details of reverse freemium design.

1. Trial Length

Trial length is important, even for freemium. For a limited time, customers will experience the magic of your full product for free. If the trial is too short, the customer will not reach the aha moment and activate. If the trial is too long, there will not be a forcing function to convert, which costs you money over time.

Product teams need to create a sense of urgency without being off-putting. One key factor is the natural cycle of using the product. Does your ideal customer use your product daily? Weekly? Monthly? Monthly-use software by definition needs at least a month-long trial. The more frequently the product is used, the shorter the trial should be.

For example:

- 4x usage per month = 1 month trial
- 1–2x usage per week = 20 day trial
- 3–5x usage per week = 14 day trial
- 5–7x usage per week = 10 day trial
- More than once a day = 7 day trial

In general it can take 3–5 serious active usage sessions of your software for a target customer to activate and get close to conversion. Your trial period is about providing an adequate opportunity.

In practice, if you have an efficient product data analytics process in place, you can plot the distribution of individual customer or cohort usage frequency to determine exactly how long your trial should be. Pick a 50–80th percentile time-frequency band and use that to inform your trial period interval. For example, if 70% of your customers use your product 3–5 times per week, start with a 14 day trial.

2. Messaging

During the trial, customers are auditioning your product for their tool of choice. Product teams will generally use a surround approach to nudge the customer in the right direction through email, SMS, and in-product assistance UI. Every single time your customer logs in to the app during trial is an opportunity to showcase something they will find compelling. The more dynamic and customized this messaging sequence, the better.

Outside the app, it's key to deploy email or SMS to send messages that will make customers more likely to return to your product. Many of the actual details of the messages will be the subject of growth efforts on your team.

3. Ending the Trial

Trial end is a distinct moment in time that represents a strong opportunity to convert the customer, especially if you have kept the customer informed (via email or text) that the trial is coming to an end. Many great product-led growth companies will make a discount offer at that time or at least announce the downgrade to the free plan so it's not a surprise. Product managers and designers should view this as a key event and focus some kind of communication on cajoling and/or informing the customer about what to expect next. Teams that opt for offering a discount should do so carefully with an eye on making an offer that doesn't impact the LTV of the customer.

If the customer does not convert, they are downgraded to the free tier. But that does not mean they are a lost sale. Free-tier customers should continue to be part of your marketing communications drip and sales pipeline. They should still have access to your free tier of support so they can continue growing enamored with the product.

Free-tier customers are arguably the greatest asset for a product-led company. Treat them as such.

4. Post-Trial

In the reverse freemium model, customers should always be retained, whether or not they convert. This means keeping customer accounts active on a free tier, and preserving the data and personalizations they have entered into your system, even if that data pertains to features they no longer have access to.

For example, configuration on a premium integration during the trial should be intact even if not available in the free tier. That way, if the customer converts to the premium tier down the road, they can pick up right where they left off during the trial. In addition, we recommend continuing any ongoing email/SMS nurture series, although customers who are completely inactive are liable to unsubscribe at a much higher rate than those who are episodic or light users of your product.

In this case, it might be worth putting them on a separate nurture list different from your active customers, so you can focus their communications on conversion vs. increased use, which would be more appropriate for converted customers.

Finally, depending on the qualification of a customer, unconverted post-trial customers can be contacted by a sales or support person to inquire about where they are in their evaluation journey. PLG prioritizes low-touch conversion, but post-trial outreach is often a good use of human resources for the purposes of learning and the opportunity to convert a valuable customer with a little TLC. Who wouldn't want to check on a sign-up from Mark Zuckerberg if he tried your product?

With freemium trials, a lot of care has to be given to selecting the right free and paid features. We'll cover that topic next.

SEGMENTING FEATURES

In a freemium model, there are features that are free and some that are paid. Finding the right mix of features on both sides of that divide is crucial to get right. In fact, you will need to make this choice for every new feature—should it go in the free column, or the paid column? And if it's a paid feature, in what tier does it belong?

The old way of thinking about trials was to be stingy with the free plan to convince customers to convert. While conversion is still the goal, being stingy can backfire spectacularly with today's customers. People will refuse to be strong-armed into paying for a product and would rather move to your competitor than acquiesce.

Great product-led teams are generous with their free tiers, providing robust features and competent support, trusting their ability to deliver on premium features for paid customers. They value their free customers, seeing them as potential energy for future conversion, and strive to make their use as delightful as possible.

Following are some principles for segmenting features between free and paid tiers.

Focus Free Features on Driving Activation

One common mistake we see with designing free tiers is paywalling the final step in the core workflow to "convince" customers to convert. We really don't like picture tools that will prevent you from using the creation you just made by putting an ugly watermark on it. This will only drive customers to your competitor. Your free tier needs to allow customers to complete the full, core workflow, thus driving to increased use and activation.

The free tier is designed for casual and persuadable customers who are still evaluating your product and comparing you to competitors. Without a free tier, products would only attract the most motivated customers who have a critical need for the workflow and are willing to pay for it immediately. Your free tier needs to be useful enough to keep non-paying customers around, giving you a chance to convince and convert them in the future. We know it works—Ezinne recently paid for Canva, a tool she had installed on her Mac desktop months ago. It had been in use for months but she suddenly needed more powerful features from it, and converted.

		Engage with free features			Monetize with paid features

Non-customers	⟶	Casual	⟶	Persuadable	⟶	Motivated
Not really affected by workflow you solve or how you solve it		Workflow is infrequent, time may not be right, affordability issues		May not be the right time, has alternatives, affordability issues		Right time for discovery, few alternatives at this price

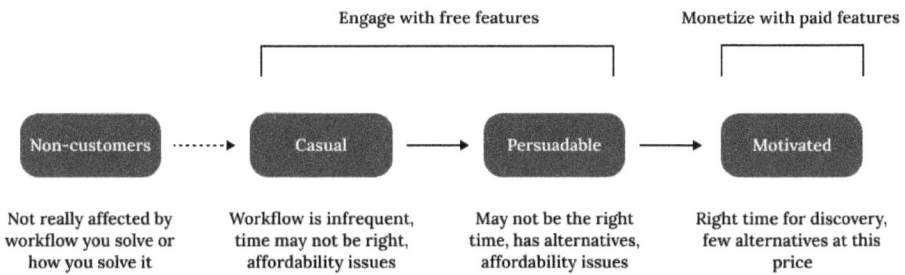

One way to intuit what should be free is to identify the smallest loops (i.e., discrete workflows) completed by 80–90th percentile of your customers that become activated. This usually represents the functionality needed to achieve that aha moment. For example, the smallest loop within Calendly is booking a meeting. Once you identify the smallest loop, experiment with how much more to add to that makes sure you're balancing generosity with profit.

Paid Features Are for Unlocking Superpowers and Collaboration

Paid features should be designed for your power users and geared toward conversion. They use your product to create real value for themselves or their company. They need more storage, support, reporting, and forecasting features. Power users are often team leaders as well. When coordination becomes critical, customers are more willing to pay.

But how do you know exactly which features your most motivated customers will value? Which will give them the superpowers they need to perform at their peak? Again, it all goes back to customer discovery and listening. Here are four principles we use regularly to inform our feature segmentation:

1. **Ensure customers can complete a core workflow in the free product.** Like we mentioned in the previous section, it's critical for free-tier customers to complete a fully functional workflow. This drives activation.

2. **Add restrictions based on scale.** Limit free-tier customers usage based on amount of output, frequency, project size, and number of teammates. For example, a project management tool may be free for a single user up to three projects, but adding unlimited projects and teammates should be paid features.

3. **Test the restriction points to preserve free tier utility.** Map out the feature restrictions you plan to put on the free version and test to see if free-tier customers are coming back despite their unavailability. For example, how much storage should you provide free-tier customers? If the restrictions are too deep, evaluating and casual customers will stop using your product. You need to find the balance of activating free customers and converting paying customers. The key question to ask is, "Are customers activating on our free tier?" The answer has to be yes.

4. **Test the conversion potential of all new features.** As you test restriction points, isolate specific features to determine which ones drive the most conversions to paid. Which have no meaningful impact on conversion rate? Most products will have one killer feature that compels their target customer to pay. What is it?

Use your customer discovery surveys to evaluate features as well. At Calendly, we discovered via survey that customers were eager and willing to pay for one key feature: the ability to connect multiple calendars. We built two paid tiers: The lower tier allowed customers to connect up to 2 calendars, which applied to most customers. The second tier supported up to 6 calendars, which appealed to the most demanding customers.

As a rule of thumb, any new feature that is not part of the basic core workflow (or a new core workflow) should become a paid feature. But don't take our word for it—test each feature yourself. We will discuss subscription points in the next section, but you can start by putting up a paywall around new features to see if they drive conversion. Features that are ineffective at conversion can eventually be moved into the free tier to enhance the free core loop.

	Free	Paid
Calendly	• Book 10 meetings a month • 1 meeting link	• Up to unlimited meetings • Unlimited links
Jira	• Unlimited projects, goals, and views • Limited automations • Limited storage	• 100x storage and automations • User roles and permissions • Higher-level security • Cross-team collaboration
Figma	• 3 design projects	• Team librarys • Dev mode • Team admin
ChatGPT	• Older models • Limited queries • Longer wait times	• Latest model • Custom GPTs • Other tools • Unlimited queries

Designing Subscription Points

Beyond nudge UI and messaging flow, freemium products have an additional weapon in their arsenal for converting customers: designing the paywall points or subscription points.

In general, a product should have all the advanced functionality in full view of every customer. However, when they try something that is in a paid tier, they receive a message explaining the value of the feature and urging them to convert within that context. Many products use a pop-up window or an info bar that slides out from the side of the UI. It should be very similar across all features (we recommend a consistent subscription component) to make it familiar, but also easy to optimize. This UI should be paired with the most friction-free way to pay for your software.

Customers should see these restricted features prominently, tempting them with advanced functionality that promises even greater productivity and super-powers—and they are just a single, painless payment away from attaining it. It's like walking through an amusement park with enticing shows, rides, and games at every turn. If you have a good data system (think of a heat map of an amusement park), it's easy to track over time which features generate the most conversions.

DATA-DRIVEN SEGMENTATION

Even though we have laid out some principles about feature segmentation, your decision-making should be as quantitative as possible.

Great product-led teams survey for willingness to pay for different tiers of functionality. It's possible to actually measure the intrinsic value of major features as part of the product research process and make individual recommendations on which tier features should be in. In fact, in a product-led company, pricing each new enhancement is a discrete step in pricing strategy (something we'll discuss more in the next chapter, Pricing for Success.) Customer discovery surveys should ask bluntly: "Would you be willing to pay for this kind of functionality? If so, how much?"

Of course, this feature-by-feature probe is very different from pricing and packaging your product as a whole, but knowing the value of each feature provides extra data points that can shape that decision.

We take a more technical look at pricing and feature segmentation in chapter 6.

DON'T BE AFRAID TO EXPERIMENT

Many product teams have an irrational fear of changing their conversion strategy. We don't think it's warranted. Feel free to experiment with the product's trial and feature segments until you have something that meets your standards and growth expectations. Early on in the journey, your early adopters are good fodder for testing different strategies, and good communication with them buys you a lot of leeway for that experimentation.

Especially early on, product teams should be free to change feature tiers to see what works. Experiment with which features are free or paid, and have the freedom to bolster the free tier and the paid tiers at the expense of each other until you find the right balance. This experimental approach will serve you well as you scale. You may need to keep track of which customers were introduced at what price, but usually that complexity can be worth it very early on. Always reserve the right to change pricing.

We also highly recommend a data-driven approach to conversion testing in addition to a qualitative approach. Many startups will launch with only a free plan in order to gather data on which features represent the best paths to activation and conversion before instituting paid plans. This is a valid strategy if you have the runway.

Remember, the hardest problem with converting customers is not designing your trial or feature segmentation, but whether you have picked a sharp problem to solve. Sharp problems by definition have potential customers willing and eager to pay—your objective is to find and satisfy them. Freemium is just a tool to attract those customers. Optimizing for conversion will not matter if the problem is not all that painful or already has beloved and viable alternatives.

Assuming you have found a sufficiently sharp problem and have a plan for conversion, let's talk about pricing.

SIX

Pricing for Success:
Balancing Profit and Accessibility

Pricing is a pretty deep practice, way beyond what is obvious. Product pricing consulting practices earn millions helping companies execute smarter pricing. Many companies want pricing flexibility—the ability to charge higher prices or roll out higher-priced products without losing customers and revenues. Or simply the ability to optimize revenue by designing a better pricing plan. Others want to gain pricing power and charge higher prices to increase revenues on their corporate whim.

As such, pricing is not the exclusive domain of product management, except at the highest echelons of product leadership. It's usually a partnership between product, finance, sales, and marketing. The key questions to consider are: How much will the market bear for your product? How should you design the different packages and offerings? Companies are usually trying to find the balance between profitability and growth for each product price package, the entirety of the product's pricing plans, and the company as a whole. In this book, we will not delve into the technical details of pricing design—it's a very involved topic and many long tomes have been written about it. However, while we will not teach you all of pricing, we can cover the basics you need to know to get started in product-led companies and startups. We aim to give a rough guide to how product managers can think about the pricing problem and how to get help where they need it.

Because pricing is complex, there are various levels to the concepts in this chapter. First, we cover common pricing models. Within these models exist common features of pricing that you are likely familiar with: things like seat-based pricing (or active user pricing), tiered pricing (feature access packages), feature bundles, quantity restrictions on packages, and more. However, fundamentally we believe there are three main common pricing models for technology companies. The other ideas can be mixed and matched within them.

SUBSCRIPTION, USAGE-BASED, AND HYBRID PRICING

Subscription pricing has been the default for SaaS companies, offering predictable recurring revenue and a straightforward value proposition for customers. Customers pay a fixed price, usually monthly or annually, to access a set of functions in the product. A good example is your Netflix subscription—you are offered three tiers: with ads, standard, and premium. While we are most used to software, it's important to recognize that even things like phones, cars, and services like DoorDash also use the same model.

For each tier, customers will use different features, therefore in this pricing model, the cost basis is based on the average usage across the customer base. Some customers will have intensive usage, while others will be light-duty customers. This is advantageous for the company as it maximizes profits across all customers.

However, subscriptions can be seen as unfair to low-usage customers who wind up overpaying relative to the value they receive vis-a-vis high-usage customers, while paying the same subscription. Despite this drawback, subscription pricing remains popular due to its simplicity and predictability for both the company and the customer.

Usage-based pricing has been gaining traction as a fairer alternative to subscriptions. Under this model, customers pay based on their actual consumption units of the software, such as the number of API calls, transactions processed, or gigabytes stored. This aligns the price paid with the value received, making it more palatable to a wide range of customers. Companies can benefit from usage-based pricing as it directly ties revenue to the value delivered, encouraging them to focus on driving customer adoption and engagement.

The downside is less predictable revenue and potentially lower revenue in the short term compared to subscriptions. For example, if your product has a lot of low-frequency customers, you should expect a lower average revenue compared to subscription-based pricing. Another downside is that the relationship between customer and companies becomes highly transactional and customers can be surprised or upset by surges in their bill, despite the fact that it correlates to their usage. Some consumption can also be depressed if customers start to actively manage their use to minimize their costs. Many companies have also found that an exclusively usage-based pricing scheme can perform worse in an economic downturn, as customers dial down their usage or stop entirely to minimize their overall expenses. In contrast, subscription-based can be more durable in those instances, especially if there is a convenient subscription pause ability.

In general, usage-based pricing is best for technology that is *essential* to your target customer and where the unit pricing is seen as fair relative to alternatives. It also works very well in high-frequency usage applications for your target audience, making it popular for generative AI products and features.

AI-FEATURE AND PRODUCT PRICING

If your company builds a feature that integrates the latest powerful commercial AI models like those from OpenAI, Google, and Anthropic, you quickly run into a pricing problem:

The more your customers use the feature and use AI word tokens, the more it costs you and thus lowers your profit. Some of these latest models are costly! Given that you want active customers, how do you derive profit?

Most companies have turned to a hybrid approach. They include some base number of tokens in their standard packed pricing, but any usage over that threshold gets charged extra. They have to carefully consider a token threshold that allows the customer productive base-level usage of their feature or product, as well as their cost basis.

We believe that eventually, AI token pricing for even the most capable AI models will be reduced to the point that AI features and product pricing can be less complicated. AI token input costs will become very similar to mundane resources like storage or CPU cycles in the cloud. This will make it possible for AI-powered features to have simple subscription-based pricing if so desired.

Hybrid pricing combines the best aspects of subscription and usage-based models. The most common approach is to offer a base subscription that includes a certain level of usage, with additional surge usage incurring incremental usage charges. This provides predictable recurring base revenue while still allowing revenue to scale with usage. Customers get the predictability of a subscription (that optimizes for average usage) with the flexibility to pay for additional usage as needed. Hybrid pricing allows companies to serve a wider range of customers and capture more of the value they provide.

For product-led companies, hybrid pricing aligns well with the goals of driving adoption and expansion. The subscription component makes the product accessible and encourages viral adoption. As customers realize value and increase usage, the usage-based component allows the company to capture more revenue from dedicated high-usage customers. This combination of predictability and scalability makes hybrid pricing an attractive choice. While the specifics may vary based on the product and market, the future of pricing for product-led companies is likely to be some form of a hybrid model.

PRICING WHEN YOU ARE A SOFTWARE STARTUP

Pricing strategy is different for startups versus established enterprises. Very early on in building a software startup, your only objective is identifying a sharp problem that's worth solving. You need to find customers who want to use your software to solve their problems and have a high intent to pay.

Product managers and product leaders should try to persuade customers to pay for their usage as soon as possible. The critical test of solving a sharp enough problem is that someone is willing to pay for it. Free-tier customers (if your pricing allows for it) do have value, as we covered in the previous chapter, but only because of their *potential* to pay in the future or their ability to *attract other paying customers*.

Specifically, though, you are watching to see if customers make your product their tool of choice. Have you built enough value so that customers can live in your software for their specific workflow? Famously, it took Figma three years of building to reach the point where it could supplant Adobe for multi-user web design work. In the early phases of a software startup, you are trying to find the *minimum feature set* that turns customers into regular and productive customers with minimal churn. Your ability to create valuable and livable tools will determine whether you can attract a million customers or if you will remain a niche product.

Following are a few more pricing recommendations for the startup phase.

Its OK to Be Free While You Establish Product-Market Fit

While we don't advocate for it, giving away your product early on can be a good way to gather customer data and learn what it takes to build a "good enough" version of your product that high-intent customers are willing to pay for. There is a lot of startup advice that says giving away your product for free is verboten; we don't really subscribe to that if there is a valid purpose like customer insight. The journey to product-market fit always takes some time, and this early customer feedback is a great resource if you can afford it. Feedback—and your ability to derive insights and improvements from it—is the lifeblood of valuable products, and early adopters have shown time and again that they can endure rough edges, especially if you give them a price incentive.

Pick your early adopters wisely. Choose from those who have the deepest need and the greatest empathy for your startup journey; those who are willing to endure the bumps and still be your biggest fans. Find early customers who will be ready to endorse and refer.

At the same time you must avoid demanding and low-empathy customers who will want you to create software around their specific workflow in ways that cannot be generalized for the customers that come later. Remember, good software is opinionated on the best workflow, only providing customization for expert users who are willing to bear the responsibility and consequences. An exception to this guideline is service-driven products. If a product is partly delivered as a non-scalable, human-driven service, especially early on, we recommend pricing early and invoicing early. This will reduce the cash burn of the wages to deliver the service. However, even when a company does this, they shouldn't lose sight of scale. As the company absorbs service delivery into software and costs go down, pricing should change to reflect it.

Start to Collect Money as Soon as You Can

While it's ok to be free, your team should feel plenty of pressure to monetize as soon as possible. Don't get comfortable. Start collecting revenue *manually* as soon as you can. Discuss pricing with early customers within feedback sessions and send them an email or paper invoice when they agree to pay. Focus on the aforementioned high-conviction, high-empathy target customers who are activating on your product.

Pushing for revenue gives your team insight about what activated customers want and at what price point. Many early customers may engage with your product, but bringing up the pricing question quickly discriminates the serious value receivers from the dilettantes. It also piles on the added *realistic* pressure of listening to and satisfying paying customers. This is something your entire team has to get good at, as long as you're in business.

It's Ok to Not Have Pricing "Hooked Up"

Early on, it's ok to not have pricing and payment automated. Invoices (really just emails with a payment link) work just fine for early customers. However, you should also set up a pricing page, even during this initial period.

Many teams get nervous about creating a pricing page ("What if we get it wrong?") but the reality is that pricing should be seen as *experimental* early on. Having a pricing page also communicates to your early customers that while the product may be free today, you intend to charge for it in the future. This sets the right expectations for customers but also gives them a chance to get hooked on the product at no initial cost (like an extended reverse freemium trial). It's also advantageous to position early free trials as essentially a *discount*.

Your pricing page is also prime real estate for learning. Consider adding a feedback form to your pricing page so that you can collect intel about pricing really early, even from those who choose not to pay. Adjust your pricing plans based on feedback and keep early customers in the loop about the changes. You'd be surprised at how accommodating customers are with this kind of transparency.

Price Competitively

So how should you set pricing early on? Because your product is new, we recommend anchoring your prices to that of your competitors and your market. Use the pricing of your competitors as a starting point and then adjust based on the differentiated value you deliver.

If you are in a competitive market (or one with many acceptable alternatives), we suggest value pricing to start. Early on, winning on price is one of the easiest battles to fight against incumbents. An exception to this rule is if your business strategy is contingent on premium positioning. In that case, value pricing may harm this objective—it's better to hold off on setting a price and charging it until you have the markers of the premium positioning you aspire to; then set the price accordingly.

PRICING FOR POST-PRODUCT-MARKET FIT COMPANIES

At the pre-PMF stage, pricing strategy is focused on maximizing customer activations and learning. As a company scales, its priorities shift to capturing market share and driving revenue.

There are three main pricing principles that product-led growth stage and enterprise companies must bear in mind:

1. Balance generosity with both access and market dynamics.

2. Price consistent with your market positioning to extract maximum value.

3. Make the pricing scheme easy to digest for your customer.

Balance Generosity with Both Access and Market Dynamics

The first principle is about making sure that your pricing scheme allows your customers to flow from *tryers* to *buyers* very easily. Tryers who fit your target customer profile should find your pricing very palatable and will usually convert to your middle-tier price point.

But growth stage and enterprise companies are defined by their ability to move upmarket or horizontally into new adjacent workflows. Therefore, your pricing scheme must also appeal to those who don't fit your ideal customer profile and may need more time to evaluate your product. This is where freemium and starter plans work well. It's important to remember that the cost of support is non-trivial for customers, so be specific and prescriptive about how much support comes with lower-tier plans.

Price Consistent with Your Market Positioning to Extract Maximum Value

The second principle is about making sure your non-freemium or starter price points/tiers reflect your market positioning and the maximum prices you want to extract from your customers. An example of a market positioning is whether you are premium service (like Apple, which prices at the high end of the market) or a value service (pricing more in line with the market and indeed offering slightly lower prices, for example, like Atlassian).

Premium prices usually go with a proven ability to be differentiated (attested by your customers) and any intellectual property that allows you to protect that differentiation. Value pricing usually comes with competitive markets and with the product offering as the insurgent entrant.

For example, Atlassian's pricing has always been value pricing. This was in line with their company value, "Don't F&%@ the customer." So when I found myself building their first new product post-IPO, Atlassian Stride, we chose a price point that was very aggressive. Compared to Slack and Microsoft Teams, which were direct competitors, we had exceptionally good value pricing. We launched with a price of $3 per user across the enterprise. This was meant to make it easy to adopt Stride as the team communications platform that could scale from the front desk to the C-suite at a comfortable price. The competition hovered closer to $5–$15 per user.

This fit our positioning well: It was a new unknown product from a well-known collaboration company with a value pricing stance for its other product lines like Jira and Confluence. In fact, that price point had a few properties:

1. It was a price increase from our older product, HipChat, to cover our high development costs for a new system.

2. There was extensive research and analysis about the prices our customers would bear.

3. It undercut the main incumbent's (Slack) pricing by 30–40%.

This neatly illustrates the balancing act involved: It *still* reflected value pricing, but we also had pricing headroom (vs. legacy product, HipChat). We exercised that slightly higher pricing option to maximize our returns.

Making the Pricing Scheme Simple and Easy to Digest

The final principle is often undervalued. While your product is scaling, it's important to use a pricing scheme that customers are both familiar with and that doesn't offer obstacles to buying decisions.

The standard pricing model in SaaS today is the per-user subscription, but hybrid pricing schemes are becoming newly popular—the challenge is presenting the mixed pricing scheme in a simple and obvious way.

While you are scaling, focus on the ability of your target customer base to easily understand your pricing, keeping a keen eye out for objections. For example, foisting usage-based pricing on customers who are accustomed to subscription-based pricing can be counterproductive if the pricing itself adds friction to the buying process. Only introduce these changes where the benefit is very high—for example, usage-based pricing can often mean significantly lower per-customer revenue in the short term, but higher conversion, a potentially good tradeoff.

If your product is often paired with other software products (for example, as part of a marketing tech stack), it's advantageous to be priced similarly to those products. If your pricing scheme is different or an outlier, it may be harder for your intended customers to pay for your tool differently, and may push them to look for alternatives. Basically, don't charge in Bitcoin when everyone around you is charging dollars, unless there is a darned good reason!

One of the most interesting things we have seen as technology companies experiment with pricing is the *one-price* company—startups will offer a singular price (paired with a free tier). This makes sense in many ways: There are no confusing price tiers to choose from. It's a simple, single, accessible price point. In reality this cleverly masks a startup's inability to invest in many features that can be price differentiated early on. So it's win-win for the startup and customers alike. Customers get a great accessible price and fewer decisions to make about products they want to use. Startups get to NOT worry about the intricacies of pricing beyond their current stage and have time to expand their offerings (and set new price points) as they grow and expand.

Start to Experiment with Enterprise Pricing

As your company scales and builds a reputation of reliable products, you will likely draw interest from enterprises looking for your services. Enterprise pricing is a different beast altogether, so scaling companies should use this early interest to experiment and refine their enterprise go-to-market approach.

We believe that at the core, enterprises are paying for peace of mind. They require enhanced security, dedicated support, custom integrations, and higher levels of availability. Meeting these needs comes at a significant cost to the provider, which is reflected in the higher prices charged to enterprise customers. In addition to the base product, enterprise pricing often includes line items for professional services, custom development, and user training. Enterprises may also require flexible billing arrangements such as quarterly invoicing or custom payment terms. Supporting these requirements requires a different go-to-market approach, often involving a dedicated sales team and solution architects.

The higher-touch sales and support model leads to significantly higher customer acquisition costs compared to self-serve models, but this is offset by the much higher lifetime value of enterprise customers. While a self-serve customer may pay a few hundred dollars per month, an enterprise deal can easily reach six or seven figures annually. Margins also tend to be higher on enterprise deals despite the higher costs of serving these customers. They are willing to pay a premium to ensure that the solution meets their needs and that they receive the necessary level of support.

With enterprise, the costs are higher, and so are the rewards. For companies that can navigate the complexities of enterprise sales, the payoff can be substantial in terms of both revenue and profitability. As we have mentioned previously, even companies started as self-serve product-led growth companies must master enterprise sales to keep growing.

Become More Precise with Measuring Your Cost Basis

Really understanding your cost basis for delivering your service can make you quite precise in constructing pricing. There are two main ways to do this:

1. The first is installing product telemetry and repeatable data analysis that makes it easy to analyze how much each customer or groups of customers consume of your product and how much it costs your company.

2. The second is analyzing the historical cost of service and finding average usage by different kinds of customers—freemium, paid, and possibly by customer profile.

This data, combined with an understanding of the lifetime value (LTV) of your customers at different tiers, produces a baseline calculation of your base profit margin and how much you can efficiently spend on customer acquisition efforts.

At Calendly this analysis helped us clarify that the LTV of a free-tier customer was net positive to a significant extent. This was mostly due to their contribution to the virality of the product: Free customers still had a high *net referral* effect. This made it easy to keep making the decision to keep the freemium plan. Without this kind of data, it would have been a guessing game.

Knowing your unit costs and LTV per pricing tier will give you a more complete understanding of the impact of specific price changes on your margins. In particular, it will give you more confidence if you adopt a value pricing strategy.

COSTS, MARGINS, AND PRICING TIERS

Effective pricing is a critical component of a successful software business. Previously, we have discussed the product-led principles of pricing, the ideas that lead your company to *low friction*, *generosity*, and *psychological safety* for your customers. However, there is another set of concerns that needs to be melded to those principles to deliver on effective pricing: the hard numbers. The union of both is the origin of good pricing.

The numerical parts of pricing requires a deep understanding of your costs, your customers, and your market. Get it right and you'll be able to maximize revenue and profitability. Getting it wrong can mean you struggle to scale.

The foundations of your pricing strategy rests on three key factors: your costs, your desired margins, and what the market will bear.

Simple Costs and Margin Math

Start by understanding your cost basis per average customer or per unit of value—for example, the true cost of delivering your product to a single customer or a single page view[17]. Let's say you have a SaaS product with the following monthly costs per customer:

- Hosting and infrastructure: $2
- Customer support: $5
- Product development (amortized per customer): $3
- Sales and marketing (amortized per customer): $10

---

17   For Typeform it was a response to a survey. For Calendly it was booking a single meeting. For WP Engine it was delivering a single web page from their infrastructure.

In this case, your total cost to serve a single customer is $20 per month. This is your cost floor—your pricing must be above this level for you to have a viable business.

Next, determine your target margin. Let's say you're aiming for a 70% gross margin, which is fairly typical for a SaaS business. This means that for every $100 in revenue, you want $70 left over after accounting for the cost of goods sold (COGS).

To calculate your base price at a 70% margin, you would use the following formula:

Base Price = Cost / (1 − Target Margin)
= $20 / (1 − 0.7)
= $20 / 0.3
= $66.67

So based on your costs and target margin, your base price should be around $67 per month.

However, you also need to consider what the market will bear. Research the prices of comparable products and understand the value your product delivers to customers. If comparable products are priced at $50 per month, you may struggle to sell at $67 unless you can clearly demonstrate higher value. Conversely, if similar products are selling for $100 per month and you're confident you can match or exceed their value, you may be leaving money on the table at $67.

Let's say after market research, you determine that $75 per month aligns with the value you provide and is competitive in the market. This would give you a margin of 73%, calculated as follows:

Margin = (Revenue − Cost) / Revenue
= ($75 − $20) / $75
= $55 / $75
= 73%

Designing Your Pricing Tiers

With your base price of $75 per month set, the next step is to design your pricing tiers. Tiers allow you to segment your customer base and capture more of the value you create.

A typical SaaS application will have three or four tiers, often named something like Starter, Pro, Business, and Enterprise. Each tier should provide increasing levels of value, whether that's through additional features, higher usage limits, or enhanced support.

Starter	Pro	Business	Enterprise
• $75 / month • includes core features • 1 user • email support	• $150 / month • advanced features • 5 users • phone support	• $300 / month • API access • 20 users • priority support	• custom pricing • single sign-on • unlimited users • dedicated account manager

When deciding which features to include in each tier, consider three factors:

1. Cost to deliver: If a feature has a significant ongoing cost (e.g., dedicated support or increased infrastructure costs), it should be placed in a higher tier with a corresponding price increase.

2. Value to the customer: High-value features that customers are willing to pay a premium for should be placed in higher tiers. This could include advanced analytics, integrations, or security features.

3. Broader market appeal: Features that have broad appeal and could drive upgrades should be placed in lower tiers to make the product more attractive. Niche features that only appeal to a subset of customers are better suited for higher tiers or sold as add-ons (a la carte).

Develop a Pricing Framework

As you develop new features, you need a clear framework for deciding which tier to place them in. This should be based on the criteria above. Features that are costly to deliver, provide significant value, and appeal to a niche audience are prime candidates for higher tiers or add-on pricing. Features with broad appeal and low delivery costs are better suited for lower tiers.

It's also important to continually reassess your tiers as your product and market evolve. What made sense at one stage of your growth may need to change as you scale.

An example simple pricing framework:

1. We're committed to reverse freemium.

2. We will have three self-serve tiers (excluding free tier) and two enterprise plans to maximize product-led adoption and connect it to sales-led adoption.

3. Low-cost and simple broad-based features go in the free tier, limited by usage and frequency, to hook customers and keep them engaged.

4. Features that provide significant workflow compression go into the top two professional plans.

5. High usage of all features require paid tiers.

6. Niche features can be bought as add-ons by individual customers. We will make them easy to discover in our pricing page and subscription points.

7. We will implement surge pricing beyond the set base subscription limits.

Remember, the goal of your pricing tiers is to provide a clear upgrade path for customers while maximizing the revenue you generate from each segment. By aligning your tiers with the value you deliver and the needs of your customers, you'll be well-positioned to capture more of the value you create as you grow.

The job of pricing is never done—it's only satisfied for a time. Frameworks will help you keep a coherent pricing strategy long term while actual prices change more frequently. Similarly, we need a framework for measuring the things that matter most to a product-led company. That's next.

Pro Edition: Conjoint Analysis

Conjoint analysis is a statistical technique we use to determine how customers determine the value of certain products and features. We provide a full breakdown of conjoint analysis, with examples, in the Pro Edition of this book. Get it at productmind.co/brpro

SEVEN

Measure What Matters:
Metrics for Product-Led Growth

Every ambition of a creative software product team has to be expressed in some kind of data metric to measure its progress. Usually they are an expression of how much customers use, adopt, or pay for certain products or features. Setting these metrics carefully is key—selecting the right metrics makes the team focus on the right activities to increase the overall value of the company. The wrong metrics can lead the team astray.

Setting objectives based on key metrics and customer data is obviously not a new idea. It is one of the oldest principles in management. There are literally dozens of methodologies and frameworks to choose what metric to measure. Some of the most common ones are the HEART framework, North Star Metrics, Pirate/AARRR framework, etc. Too many companies get bogged down in dogmatic arguments about which system is best.

Assuming you have a good enough goal-setting system, what's more important is how well they are used *to provide visibility to the squads* and *drive change*. Where product-led companies differ from traditionally managed companies is they prioritize holistic and complete datasets that are accessible by almost anyone in the company. It's crucial that each person at the edge of the company can use this kind of data to optimize their goals, therefore empowered transparency and data-sharing are often core values in product-led companies.

Remember, one key aspect of being a good product-led company is being responsive to customer feedback. The goal is to delight and empower customers in such a way that purchase or license expansion decisions are easy to make and justify. Responsiveness to customers is amplified when virtually everyone on any team not only has access to clear objectives, but also critical product-usage data that tracks how well the objectives are progressing. Everyone on the team should be able to easily ask and answer second- and third-order questions about the data that informs progress to those goals. It's important that they can easily satisfy their curiosity about any kind of usage or correlations that are not immediately obvious. The distance between curiosity and data-informed decision-making should be measured in minutes.

Imagine a product manager has a question (say during the morning shower) that might inform a crucial decision, and they are able to open up a data analytics tool like Amplitude, and search for a valid answer to their question before their 9 a.m. standup. A situation like this means that data-driven decisions can be made very quickly hundreds of times a day, which adds customer-centric accuracy to your creative team.

It's hard to understate how much acceleration can compound from achieving this kind of fluidity with product-customer data. The best analogy is compound interest: small, daily investments creating incremental improvements that eventually add up to exponential growth. Consider that the alternative is either very little insight into product-customer data—which is the state of a vast majority of product teams—or mediated through data analysts who may take days or weeks to answer basic correlation questions that are needed to make good decisions.

DEFINITIONS: METRICS VS. GOALS VS. OBJECTIVES VS. TARGETS

Metrics are the data dimensions we want to measure. For example, we may choose to measure customer activations, monthly active users, or friend requests sent.

Goals, objectives, and targets are all synonymous with the specific threshold you are trying to hit. You may set the goal of 100,000 activations, 700 million monthly active users, or 7 friend requests in the first 7 days.

We will use the terms "goal," "objective," and "target" interchangeably. We don't believe in dogmatic definitions, and don't want you to feel boxed in either. Use the term that feels right for you.

PRODUCT-LED METRICS

So what kind of data should be tracked, measured, and targeted? A good measurement system can zoom in on one individual and also paint a picture of cohorts of thousands or millions of users across multiple features and workflows. Many data tools can dive into questions like:

- How many people/cohorts use this product?

- How does that work in certain time frames?

- Which features do they use most?

- How often do they use it?

- What is their path through the product?

- And crucially . . . Where are we losing them in their journey through the product?

The last question can be answered with an interesting visualization called a Sankey chart. Most modern product-analysis tools have this built in. It's a very important tool for seeing whether customers are having the product journeys that product managers and designers expect them to have, and if not, how might they solve the frictions still present in the product experience.

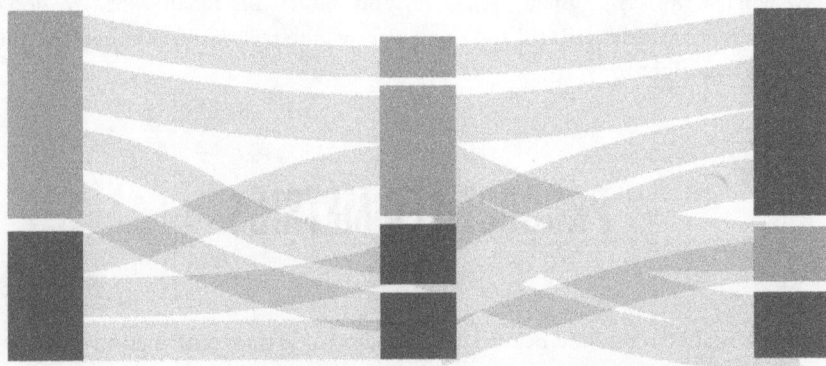

Let's dive in to some more of the types of data present in good product-led data operations:

Customer/Cohort Data

Customers should be identified on an individual level. Note that for customer privacy reasons, product teams prefer not to use personally identifiable information to do this. Using an anonymized ID number will suffice. Nevertheless, technology companies should track individual customer journeys from the time they land on your website or app, through activation, conversion, and until they churn and beyond.

Tracking individual customer journeys allows for two main advantages:

1. Customers can be tracked individually but also organized into grouped cohorts. For example, you can keep track of customer behaviors along the following example dimensions:

    a. All customers who signed up last month

    b. All customers who claimed they were sales reps that became active in the last six months

    c.   The person who signed up on Friday morning

    d.   All people that signed up last week who were students

2.   Customer identification can be combined with other data elements to really ask and understand complex behavior. For example:

    a.   Which kinds of customers subscribe fastest (i.e., mixing cohort data with subscription data)?

    b.   Which kinds of customers give the highest net promoter score?

These combinations can be quite powerful to help tease out important unobvious data correlations that might help teams make better decisions.

Product Usage Data

Product usage metrics cover how customers are actually using your product—their level of activity, what they use and what they don't, and how they journey through your product. In principle, product usage data is simple—it tracks what each customer does (anonymized, of course) and you sum this up across multiple customers if needed. It also often captures sequences so you can visualize Sankey maps of how cohorted customers navigate through the feature or application.

Product usage data should be supplemented with screen-recorded customer sessions. This is very useful for product managers and designers who want to watch how customers use new features so they can visually see dead ends and cul-de-sacs that were hard to anticipate. Usually these visual session replays can be very valuable when optimizing key experiences like onboarding or experimental new functionality.

Financial Performance Data

Financial performance data about your customer's selection of subscription levels and usage is crucial to understanding how product and customer data interacts with subscriptions. When a company knows what their longest-tenured, most profitable customers look like compared to their least-profitable customers, they can make crucial decisions on how to improve the product, improve marketing, and grow more rapidly.

Subscription metrics may not quite capture all the motivations of your customers in the same way a good sales rep can for a large account, but they don't need to. Subscription metrics capture who is voting with their dollars to maintain access to your software or technology. This tells you who values your product most.

Once customer, product, and financial data is sufficiently visible in an easy-to-use data analytics regime, it's time to use them to isolate the most important metrics and then set good objectives for your team.

EXTENDING AND CONNECTING DATA IN USEFUL WAYS

Sometimes there is extended data that is possible to gather about your customers. For example, it is useful to tie product feedback from customer listening methods into your data system.

Another example: Connect your Net Promoter Score (NPS) data to customer ID so you can see what kind of customer is rating your product highly (so you can see early which customer personas are registering product-market fit)—or even the average feedback sentiment of a certain cohort. Lots of data can be collected disparately and may actually come from multiple isolated tools. However, data is much more valuable when it can be combined in interesting ways, so connecting disparate data often yields additional insight that is useful for observing the truth about customer behavior and product success.

In general, when you can connect customer sessions to a single-customer ID key for all your qualitative and quantitative data gathering, do it! The ability to get really great analysis using that consistent thread through all your data, now or later, is almost always invaluable.

LAYERED OBJECTIVES: ALIGNING YOUR TEAM

Setting good team metrics and objectives seems simple at first glance, but the process is complicated by the fact that individual product teams don't exist in a vacuum. Their metrics and objectives must be connected and ladder up together to help the company reach its highest-level goals.

Metrics do more than just keep score—they are also often a system for motivation for a company and its teams. Tying team and personal success (financial reward, promotions) to the attainment of certain metrics is important to create incentives for everyone in the company. Careful selection and goal-planning is critical to drive the right incentives that lead to company success.

The sections that follow walk through how to think about setting the right metrics, connecting from the top of the organization down.

Start with Your Company's North Star

Many companies use a single metric—usually called a North Star goal, a.k.a., North Star—to guide an organization. A North Star goal contains a *single metric* and a *target* for that metric over a specific time frame.

A single metric is psychologically useful for creative organizations because it can be used to galvanize employees around a clear target. It gives people a simple scoreboard to focus on maniacally and is refreshingly clarifying for everyone involved. Company leadership then sets a target for their *North Star metric*, making it a *North Star goal*.

An example for a North Star goal from Oji's time as Head of Product at Twitter is the focus on maximizing *daily active users* (DAUs). Each year his team would set a new target: e.g., 300m DAUs.

There are a few recommendations for selecting a good North Star goal:

1. The North Star metric must represent customer value. For example, "active customers" is a better North Star metric than recurring revenue because the former is tied directly to customer success and value, while the latter is a second-order outcome from those customers finding value. Getting the first thing—300m active customers—pretty much guarantees the revenue if the business model is well designed. However, one is more customer-focused.

2. The North Star metric and goal must express something fundamental about the vision and mission of the company. E.g., "1 billion PCs bought" is an extension of "a desktop in every home." (More on vision and mission-setting in chapter 14)

3. It's best if the North Star goal is a simple whole number vs. a relative number. For example, "1 billion PCs sold in 2024" is better than "grow PC shipments by 25% in 2024." In this example, Company X sold 800m PCs last year. Simple whole numbers contain all the information needed vs. hunting for another baseline number to understand the goal. It's much harder to think in relative terms, and baselines can be fudged or be litigated.

4. The metric must be a leading indicator that predicts future performance rather than summing up past performance. E.g., "number of deals signed monthly" is a leading metric compared to "size of last quarter's revenue."

5. The North Star goal should be actionable and under the control of the team (or what is the point?). For example, "reach 5 million active customers" is more in your control than "win an industry award."

When a company is aiming for a singular North Star goal, leadership needs to be sure it is the absolute right target, expressed in the right way to efficiently create motivation and momentum.

Work Backward from the North Star

The North Star goal is a guide for the company, but it is not a roadmap or even a strategy. Just like intrepid travelers across history who used the North Star to navigate, product teams still need to pay close attention to the terrain around them. With your company's primary ambition set, company leadership must then translate that goal into more manageable subordinate metrics and objectives that, when combined, will deliver the company to the promised land.

Many product leaders are advocates of "layered" objectives or OKRs because they force alignment up, down, and across the organization. We generally agree with this approach without being too dogmatic about strict definitions and rituals. We simply advocate for a set of layered or "tiered" metrics and goals that guide every team toward the company North Star.

Let's talk through an example of how layered metrics work. This example comes from Oji's tenure as Head of Product at Twitter (the situation is real but the exact metrics and objectives have been changed):

Twitter's North Star metric at the time was Monthly Active Users (MAUs), and their goal was to reach around 700 million active customers/month. Their MAUs at the time were around 450 million active customers/month. For reference, Reddit had 300 million MAUs and Facebook had 1 billion.

The North Star goal of 700 million MAUs was set by company leadership, including Oji. But what are the inputs required to reach that goal? Oji and his product leadership team determined three strategic Tier 1 metrics:

- Active Producers: The number of Twitter users that created content every day (less than 10% of Twitter users create the vast majority of all content on the platform)

- Active Engagers: Users who interacted with content on the platform

- Active Ad Viewers: Users who spent enough time on the platform to view an ad.

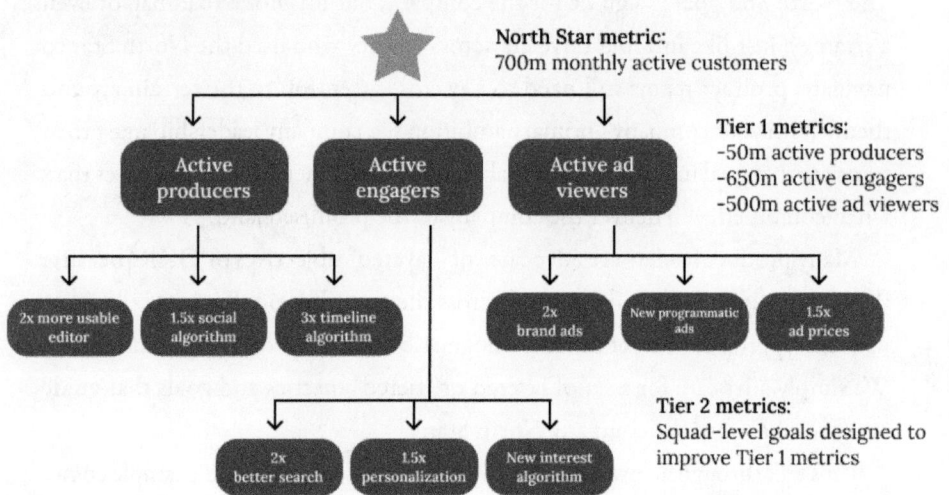

North Star metric:
700m monthly active customers

Active producers   **Active engagers**   **Active ad viewers**

Tier 1 metrics:
-50m active producers
-650m active engagers
-500m active ad viewers

2x more usable editor   1.5x social algorithm   3x timeline algorithm

2x brand ads   New programmatic ads   1.5x ad prices

2x better search   1.5x personalization   New interest algorithm

Tier 2 metrics:
Squad-level goals designed to improve Tier 1 metrics

Oji's team set a target, or objective, for each of these strategic metrics: 50 million active producers, 650 million active engagers, and 500 million active ad viewers. Each of these metrics was a crucial piece of hitting the company's North Star goal. Different product groups or divisions (if your company is large enough) would be responsible for each Tier 1 goal.

Leadership Selects Metrics; Product Teams Set Objectives

Once the Tier 1 goals were assigned, each division continued the process of alignment. They asked themselves, "What needs to be true in order to get to 50 million active producers?"

From that question they developed another set of metrics and objectives for each Tier 1 goal. This process created a layered set of metrics and objectives. In Twitter's case, each Tier 2 metric was assigned to a cross-functional, semi-autonomous team called a "squad" (more on squads in chapter 9).

Ownership of these metrics is crucial. The leaders of each division should choose the Tier 2 metrics, but the squad leaders (usually the product manager) should set their target or objective. As the folks closest to the problem, they have a keener sense of what is feasible if the team really pushes.

Imagine an army from antiquity planning their trek across the region to seize the lands of a rival. The emperor sets the North Star goal: Take over the land to the North. The general of the army then says, "Deploy 1,000 infantry battalions and 200 cavalry." But when the mission begins, it's the lieutenants—the leaders of each battalion and cavalry unit—who lead the march to the North. The "boots on the ground" are responsible for overcoming any obstacles in their path and seizing opportunities along the way. The emperor sets the mission, but the lieutenants set the best path to get there.

Back in the (mostly) civil world of technology, it's the product division leaders and PMs that set the best path forward. Given the North Star goal and Tier 1 objectives, the "boots on the ground" determine the Tier 2 metrics and objectives that are most likely to lead to success.

Each metric and its objective should have a simple narrative about why it matters and how it connects to the North Star. That way there is confidence at every level of the organization that everyone is rowing in the same direction. The logic of the plan should be straightforward. Metrics that lack simplicity are at least one sign a team needs to iterate on their metric choice.

Metrics and Goals for Mature Companies vs. Startups

Usually a laser focus on the North Star goal and its sub-metrics/objectives is appropriate for products that are mature and where the goal is incremental improvement.

However, we find that market expansion efforts that need new features or more generally, new experimental products and startups, should initially have a less-strict metrics regime than mature product lines. For mature companies, this is especially true if the target customer for these new efforts is different from the current target customer definition that product teams truly understand.

Attacking new workflows and target customers implies a period of learning: time to deepen an understanding of the customer, hone in precisely on the product or features that will solve their critical workflow problems, and generate revenue— the process of achieving product-market fit.

Firm North Star metrics and goals imply a firm grasp of product-market fit. So, early on in the search for PMF, North Star metrics are not as appropriate and goals should be oriented toward discovery and experimentation needed to lock in on the metrics that will have the biggest impact on a new business. Your appropriate North Star needs to be *found* first.

In the specific case of new market expansions in mature companies, these new incubations are often unable to deliver immediate impact to the North Star goals or metrics (they start out small and experimental). Entirely new incubations are akin to new startups. The associated product or features often begin with less functionality than is needed for a mature offering, so the team can test, learn, and build a more successful set of solutions. Usually the metrics that inform those early phases are geared differently.

Here is how to view and measure new efforts through the incubation period:

1. Have a clear understanding of how this new incubation is connected to your mission or your strategy. For example, if the goal is monthly active monetizable customers, how does a new feature fit into that? At what scale can it be material? In what way? The linkage to company mission and strategy should be apparent.

2. Set PMF goals for the initial incubation. These should include goals around adoption and early adopter happiness or enthusiasm. In addition, they should include an understanding of unit economics and how they change with this new offering.

3. Map the path of progress from an MVP (minimum viable product) to an SLC (simple lovable & complete product) and beyond. As progress is made, revisit the goals that are needed to measure incubation maturity until you can see product-market fit, i.e., you are converting target customers consistently and retaining them.

4. Product-market fit is usually the point of maturity where new feature sets should align to a more rigorous metrics/goals regime.

If step 4 is never achieved, then the incubation is a candidate for cancellation and not strategically aligned. If it's a startup, it may be time for a pivot.

Applying mature product metrics to startups and new incubations is a recipe for measuring the wrong things and may lead to premature product death. Incubations need time to become good long-term contributors to the overall objectives.

DEMOCRATIZING MEASUREMENT DATA

The systems and tools used to examine and analyze data should be simple, straightforward, and available to anyone in the extended product, engineering, design, and research organization[18] unless some kind of legal access requirement prevents it.

---

18 Consistent with what we have been saying, this should include product marketing, support, customer success, along with core engineering, design, product management, and data analysts.

The more data-informed your product teams are, the more correct decisions they can make, and the more substantive conversations they can have about critical questions. Eventually this contributes to a culture of asking the right questions and coming to the right conclusions on autopilot—a state of flow that is very powerful for a product organization. In this state of flow, almost anyone on the team can formulate a critical question and have it answered in minutes or hours, instead of days or weeks. Answered questions lead to good decisions, and good decisions stack up to a *high-throughput product team.*

As such, it's crucial that the main product data analysis tool be self-serve by anyone on the team—just log on, articulate questions, and start to get answers. Product leaders should want everyone in their organization able to quickly dive into relevant data and find answers to interesting questions. In particular, PMs, designers, and engineers should be required to be highly data literate and able to use available data analysis tools well.

There are many great tools for product analysis available today. Choose them for their simplicity of deployment, completeness of information, and intuitive analysis. There is usually a difference between the general data tool set and the data science tool set. Don't feel the need to use the same tool for both. Often data scientists are using custom tool sets to find hard-to-discover correlations while PMs are simply trying to understand usage, conversion correlations, or observe how specific cohorts are performing relative to new product features.

DATA INTEGRATION IS CRITICAL

For any tool, the data integration layer is a key element of deploying a system that is truly self-serve, but also complete. The data integration layer is how disparate kinds of data get connected and made sense of.

In a product company, data can come from different places and at different times. For example, from the key listening channels we covered in chapter 2, from a customer support tool, from a sales tool, from a product feedback tool, etc. Consider a product team that has put in their first product data gathering tool, and then added a session tracking tool six months later. They then build a mobile app and want to try to correlate mobile and web sessions from the same customers. Then a couple of years down the road, they add an automated NPS (net promoter score) survey engine. Without an integration layer, all these data systems would have to be custom knit to each other to make any sense.

With an integration layer tool, as new data comes along, you plug it in seamlessly. It's optimized to allow you to keep adding new data sources from customers or the business, indexed by customer accounts. As long as key data columns are consistent (like the customer ID, for example), new data sources pointed at your integration layer should increase your ability to draw even better conclusions from the existing data you have. In our example, you can check not only the average adoption of a cohort of a new feature but their average qualitative satisfaction with it, which can be very probative depending on what you're looking for.

By making data analysis tools a key part of your self-serve product culture, you can make everyone in product and beyond be more customer-centered in making product decisions.

WORKING WITH DATA SCIENTISTS AND ANALYSTS

Most startups can't afford data scientists or analysts. Data-intensive needs only arise from either data-centric startups, or with scaling companies of significant success, which makes data analysis worthwhile. In the early days, the best they can do is find really data-literate employees who can also analyze data really well. This is typical of many roles at a startup, so this is not surprising.

As a business grows, data needs grow significantly in two ways. First is that available data explodes. Companies can and do collect an enormous amount of data if they want to. Even if the data is not looked at initially, it's always possible to do historical analysis, so if data can be collected safely and be protected (obeying the law and maybe even more—hewing to ethical considerations), it likely should.

Secondly, as engineering teams build new functionality, they may add more tables to an existing database or seek to try new, more appropriate database systems as they realize what is needed. Most of this can be hard to anticipate.

A good product-led company thinks about data practices and analysis early. And the first line of defense is the data analysis tool set mentioned above that is meant to democratize data curiosity in your teams and help with experimentation and decision making. There are maybe 70% of product metrics questions that should be very easy to self-serve by anyone on the product team through a competent data analysis tool. For example, exploring which customers activate and convert should be easily accessible, as well as critical related ratios (e.g., sign up to activation ratio).

However, when systems are not connected and harder analysis needs to be done, dedicated data scientists are required to make sense of customer usage and financial data. Some examples of this kind of analysis include understanding the lifetime value of your average customer compared to the specific LTV of the kinds of customers who opt in to certain plans. Another example is figuring out the detailed firmographics of the kinds of companies you should target by looking at the customers you already have.

When Oji was at Twitter, he discovered it took serious data analysis to understand how and where informal communities were on the platform. It was not obvious from looking at Tweets; you had to look at topics and follow graphs in a sophisticated way. Oji's team also worked with data scientists to set the goals that the Conversations team needed to hit to materially move the North Star goal of the company.

Remember, the first line of defense of a data-informed team is democratized data analysis tools—easy to use, fast, and reliable. Most people in engineering, product, design, research, data, and sales (whom we collectively call The Shipyard, which we cover in depth in chapter 9) should have access to it and should be educated on how to use it. *Roughly 70%* of data demand on your teams should be *self-serve*.

The *other 30%* should be harder questions that are not obvious or maybe even possible to extract from the front-end data analysis tool set. Complex questions may not have a precise answer and need correlation between product data, financial data, and external data sets. For these, you likely need dedicated data scientists and analysts to help product teams have a shot at understanding these hard-to-find correlations.

Teams that work like this will have a few advantages:

1. Higher use of data tools by more people in the organization with attendant data literacy

2. Fewer high-powered data scientists/analysts needed to run the business

3. A natural understanding of high-value/low-value analysis at the appropriate level.

Product-led companies know that data is the lifeblood of their organization and build instrumentation to make it accessible and useful for everyone.

EIGHT

Engineering Virality:
Short and Long-term Growth Strategies

The speed at which a product-led software business grows is closely tied to its virality—how quickly it spreads within its intended customer base without applying marketing resources. Typically a software business that's not in a winner-take-all market starts to grow slowly and then reaches a peak of adoption where name recognition and the application of demand-generating marketing can drive reliable growth. This is roughly a baseline, but virality will generally speed the business outcomes significantly—akin to a multiplier effect.

Virality is customer-assisted marketing that grows exponentially. Its key ingredients are how revolutionary your product is (i.e., the level of augmentation it provides), the quality of your UI, and how well it facilitates social sharing. Viral products create net referrals from customers who come in contact with it.

Software used together by a team, or social software products, have an easier time being viral because your customers use them in concert with others. As a result, bunches of people can spread the word about a great product rather than single users of the product; a neat amplifier effect. This is the difference between a social note-taking app and a personal single-user note-taking app. One is a great personal product that you can brag about. One naturally invites collaborators by its function and we can all brag about it together—to, presumably, better overall viral effect.

When most people think of virality, they often think of cunning viral loops or product gimmicks that drive referrals. We define these tactics as *synthetic virality*, a type of growth that is earned through product awareness but not sustainable for the long term. Like all social trends, synthetic virality has a short half-life. It's certainly possible to engineer this kind of virality, and many great product marketers do, but experience tells us that *durable virality* is much more valuable in the long term. It relies less on clever gimmicks and is rooted in strong product fundamentals. Let's dive into durable virality first.

DURABLE VIRALITY

It's clear that some products are more viral than others. It's not just because of clever marketing or going viral on social media (though that may be part of it). Some products are enduringly viral, meaning they continue to grow exponentially long after the "viral" moment has passed. Think of legendary products like PayPal, Slack, and Figma. These products have *durable virality*. Unlike synthetic virality, durable virality is built *into* the product (not applied afterward) and has a long-lasting impact.

How do you build durable virality into your product? There are several strategies in play.

Redefine the Category

Product management has its roots in product categories. When P&G created the first "brand men" to manage their growing product line, they organized the new function based on category: hand soaps, shower soaps, cleaning supplies, etc.

Categories are still the most important factor in building a dominant product. Products that reset expectations for customers will quickly become the category leader and remake the entire category in their image. This is easier said than done—redefining a category means creating a product so different and better that all other products (especially the incumbents) are compared to it. No small feat, but it is the most important step to achieve durable virality.

There are numerous examples of category-defining products and how they create virality:

The first one is the iPhone. At launch, it commanded intense consumer loyalty, and not just because of Steve Jobs's ability to distort reality of those around him. It was because it was the first big paradigm shift in phones for a decade. People lined up for days to get one, and even switched carriers (no easy feat in itself) to AT&T, then the exclusive iPhone carrier.

More recently, Uber revolutionized ride-hailing, a product category that has existed in one form or another since the late 1800s. The discrete tasks of hailing, sharing directions, and paying were all compressed into a single, superior workflow. This was a revolutionary product and earned its durable virality that exists to this day.

How Do You Redefine a Category?

Category redefinition requires more than just product improvement—it demands a fundamental shift in how customers interact with and think about the solutions we want to deliver into the category.

The cornerstone of this transformation is developing deep customer understanding through various customer discovery techniques, including ethnographic research, domain expert integration, and comprehensive workflow mapping.

- This intimate knowledge becomes the foundation for leveraging three additional strategic dimensions: identifying and utilizing technological inflection points (like Uber's combination of GPS, smartphones, and digital payments),

- Reimagining customer interfaces and experiences (as demonstrated by Figma and Calendly),

- And timing market entry effectively.

The timing of market entry involves analyzing three key vectors:

- Technology trends (emerging possibilities and falling barriers)

- Industry trends (evolving customer expectations and regulatory changes)

- Macroeconomic conditions (budget availability and broader shifts like remote work or AI)

While companies don't need to perfect all dimensions—Uber succeeded primarily through technology leverage, while Figma won through interface innovation—success comes from identifying which combination offers the best path forward, always guided by deep customer understanding.

While perfect market timing isn't possible, maintaining awareness of these dimensions while building customer knowledge gives companies the best chance to build a category-redefining product.

It's easy to chalk up category-defining products to "great marketing" or "good timing." Hotmail, for example, is often cited as the premier example of hacked (or synthetic) virality because of the promotional tagline they added to the end of each email. Yes, this feature spread the word about Hotmail rapidly and led to viral growth, but that's not the main reason why they succeeded. The world of email was pretty terrible in 1995. Web browsers were just becoming popular and most still came bundled with a POP3 email client. POP3 worked by fetching your mail from the mail server over its specific protocol. Unless you set it otherwise, it would deliver the mail to your local inbox and delete it on the server. This meant you had to use the same computer all the time to maintain a coherent chronology of your mail. Those who used multiple computers (e.g., a desktop and a laptop) had to be extra careful not to bifurcate their email records. Hotmail solved this problem by never really deleting email on the server and providing a consistent interface to all your mail that could be accessed from anywhere. It circumvented the need to dial in to your modem and wait for your email to download, which was a feature of early email clients.

In other words, Hotmail was a revolutionary product that redefined the category of email. The synthetic viral hook was just a cherry on top.

The key to durable virality is building a uniquely valuable product. It's infinitely more valuable than clever hacks or viral marketing tactics.

But that's not to say messaging and perception aren't important as well.

Exceed Expectations

We encourage product managers to deliver a product that exceeds expectations. The biggest challenge to doing so is time. Exceeding expectations, by definition, takes more time than absolutely necessary, and many companies feel they can't spare a single day where they aren't shipping. This feeling is especially acute in highly competitive markets where time to market appears to matter, though it matters much less than most companies realize.

However, there are much more important considerations than time to market. The main challenge for any company is identifying a sufficiently sharp problem and validating if your product solves the problem well enough for customers to pay for it. Once you find this out, your focus should be on product quality—building what this core set of customers love. There is no virality without this foundation.

PayPal exceeded expectations by stumbling on a surprisingly sharp problem: eBay powersellers getting paid for their auctions. eBay launched without a payment tool of its own, so buyers and sellers exchanged checks or even cash via mail. This obviously led to a wide range of headaches, from slow payments to bounced checks to fraud. PayPal, which originally aspired to beam money across PalmPilots, had created an email payment feature as a backup to its main Palm product. eBay power customers discovered this tool and it stuck immediately. For this niche market, getting paid via email was so much better than any alternative that the community raved about the product. Then PayPal went above and beyond by building a world-class customer support organization, which kept them a step ahead of competitors (including eBay's own native payment app).

Even products that redefine their category will eventually dull and lose virality if the underlying product promise is broken once too often. If you've followed the principles of customer discovery, customer listening, and simplicity, you should be well on your way to nailing a quality product.

Build Team Software vs. Solo Software

Software that is designed to be used in concert with your work team or your friends is inherently more viral than those designed for personal use. It's simple math: If a product is best when used with 5–7 other users, customers will naturally invite their teams and colleagues to join them. Each new account brings multiple new users as opposed to just a single seat.

Besides communication tools, most software has traditionally been designed for a single person. But team software has come a long way. Think of the difference between using Microsoft Word in 1995 vs. using Google Docs to collaborate on content in 2024. In collaborative work environments, the accessibility and real-time editing of Google Docs significantly increases productivity. Teams no longer have to share a single file or keep track of the latest version. The more people on a team using Docs, the faster it will be adopted by everyone else.

Turning solo software into team software was the foundational insight of Figma, the $10 billion design platform that Adobe tried to acquire in 2023. The Figma team worked for several years to nail their multi-user experience, confident it would be their defining feature and major competitive advantage against Adobe. After launching in 2016, Figma grew rapidly to steal a significant segment of Adobe's market: web and app design teams that needed to collaborate on every project. (Note, however, that Figma still has a strong solo-player mode for freelancers and customers who want to test the product before sharing with their team.)

Easy to Share

Durably viral products are easy to share with colleagues, friends, and social networks. Not only should it be easy to invite teammates and colleagues (as we covered in chapter 4: Onboarding and Activation), but the work output should be sharable too. A data analysis tool that produces useful, stunning graphs is essentially a content marketing machine for the company. Every time a user shares a work artifact with another person, awareness of the product grows.

Shareability comes in many forms:

- Invitations to teammates
- Visually appealing content artifacts
- Social sharing functions
- Ability to tag other users

Every software team should work to make their products more shareable. The best modern example of shareability-driven virality is the Apple Ecosystem. Airdrop and iMessage are two apps that make sharing between Apple devices completely seamless and, for many customers, indispensable. Apple has engineered such strong ties between its products and users that their customers often feel a visceral dislike for non-Apple users.

Easy to Try, Easy to Buy

As we discussed in depth in chapter 3, the faster a customer gets to the aha moment with your product, the more likely they are to become an active user and paying customer. Then it's just math: More converted customers means more chances for them to share their delightful experience with others.

HARDER TO TRY, HARDER TO BUY

1 million
customers ←   → 200k converted customers

1 million
customers ←   → 400k converted customers

EASIER TO TRY, EASIER TO BUY

If a product can make their activation and onboarding process particularly seamless (especially compared to direct competitors), that product will be even more viral. A great customer experience will always be worth sharing—and the more customers a product has, the more chances there will be to share.

Sweat the Entire Customer Experience

Customer experience touchpoints beyond the product can also add to (or detract from) the durable virality of a product. No matter how well-designed the product and its UI, customers still occasionally need to talk with support. Interactions with customer success, customer support, and sales are make-or-break factors in the overall customer experience. Each of these touchpoints can become shareable moments—for the benefit of your company, or to its detriment.

Customer support is particularly important because the small subset of customers who interact with support tend to be highly involved, emotionally charged, and quick to share their experiences online. Fanatical support can be a real difference in net referral and churn prevention. There is something about support reps who really listen and solve your problems in short order and with good humor that inspires loyalty. Indeed, customers who need help and get it can often be more loyal than those who don't ever need it at all.

Calendly had an obsessive focus on delivering excellent customer support, which contributed to their virality and growth. Support was the largest organization in the company for a long time and the team was extremely personable and responsive. Oji's product team had a bi-directional meeting every two weeks with the support leaders to track the top ten customer issues, which the product team worked on relentlessly until they were solved. A lot of credit goes to Tope Awotona, Calendly's founder, for investing the resources needed to keep the team fully staffed and productive through the company's rocketship growth. There were many instances where customers were retained simply because of contact with the excellent support team, and then went on to rave about it online.

Product is the most important part of the customer experience, but every part of the organization—from product, to sales, to support, to the C-suite—should be aligned to deliver a seamless experience that makes customers want to share with their friends and colleagues.

SYNTHETIC VIRALITY

Durable virality is built into your product and customer experience. Then there are tactics that you can apply to your product to supercharge growth—if only for a brief period of time. We refer to these growth hacks as *synthetic virality*. These are tools that can enhance durable virality but will likely not be a good sustainable foundation for customer referral and marketing. Companies should not chase synthetic virality until they are confident their product has durable virality.

If durable virality is the rocket fuel for your product, synthetic virality is the afterburner. Let's look at a few of the general strategies.

Social Sharing

Even when social sharing is not a core function of a product, adding it can often make it more viral. The key is to create a social media–friendly output that your customers will be excited to share. Consumer software is great at this—games like Wordle and language learning apps like Duolingo create scorecard assets that are perfect for social sharing. More business software should take this page out of the consumer app playbooks. Any time a product gives its customers a chance to brag is a win-win for the customer and the company.

The best recent example of social media virality is Notion, the do-it-all platform for docs, wikis, project management, and workplace collaboration. Notion primarily functions as an internal team for freelancers, but any page can be published to the web and made public. Some small companies and freelancers use Notion as their primary business website. Every time a Notion page is shared on social media, another person is exposed to the platform.

Zynga is another example of a company built on social sharing virality—and then perished because of it. Zynga's games like FarmVille thrived on Facebook in the 2010s, reaching its peak with 32 million daily active users. Growth was driven by its insanely effective viral marketing via Facebook—users earned points, badges, and virtual resources by inviting friends and trading goods with them. It became a global phenomenon—and a nuisance to non-players. Facebook eventually throttled how much FarmVille players could post on Facebook, and usage quickly dropped off. The game was shut down for good in 2021 when Adobe and Facebook stopped supporting Flash.

Like all synthetic virality, growth from social sharing should not be considered eternal. Eventually, trends and platforms change. Zynga isn't a warning to *not* use social virality. By all means, if you can get it, go for it. Just don't bet the farm on it.

Referral Programs

Referral programs are a way to pay customers for sharing your product with non-customers. When new users convert, the referrer gets a bounty from the maker of the product. The reward can be a company perk, user credits, or straight cash in the form of gift cards. Companies with strong brands often use branded merchandise to incentivize their most evangelical customers.

There are many good and bad ways to implement referral programs. Good referral programs are predicated on honesty. The company implements a verifiable tracking system so people get rewarded for legitimate referrals. Receiving their rewards should be fast and simple; the only thing worse than making no money is being owed money, so companies should pay referrers quickly. Referral codes should be easy to share, and crucially, well-designed. No serious person wants to spam their network with clickbait-looking links. Finally, the team running the referral program should not be bogged down by management overhead, or else all the previous points in this paragraph become more difficult.

Bad referral programs have the opposite characteristics: slow cycle times for good referrals, a lot of management overhead, hard-to-share links, and are hard to track. Few potential referrers are desperate enough for cash to put up with a poor referral experience. As a rule of thumb, companies should assume referral rewards are icing on the cake for a customer who already loves their product and would probably have shared it anyway. In that sense, the referral program is just a nudge to share, not the sole motivator. Therefore, the program must be simple and well-designed—just like the rest of the product.

Product managers tend to underestimate the overhead in managing referral programs and overestimate their effectiveness. Use with caution.

Branded Output

A very common viral tactic is for a company to put their logo on the work output for free-tier (or even low-paying-tier) customers. This specific version of social sharing is particularly popular in B2B SaaS. This is what Hotmail is best known for: branding each email sent by its users (a tactic that was later copied by Blackberry, Apple, and other email providers like Superhuman). Another example is Calendly, which features its logo prominently on each scheduling page. Paying customers can opt to remove Calendly's branding, but free-tier customers cannot. This is widely considered a fair tradeoff for a freemium tool.

Unexpectedly, a really good example of branded output in the consumer space is TikTok. It is surprisingly easy to export video from TikTok to share to other social channels—especially compared to other social networks like Twitter and Facebook, who only let you export links to the video. However, at the end of each exported video is the TikTok logo and a branded sound. As these videos circulate, each one becomes a branded ad for the platform.

Viral Payoff Pages

Viral payoff pages are used to capitalize on transactional software experiences that involve two or more people. For example, after a customer signs a document using DocuSign, or schedules an appointment using Calendly, or completes a Zoom call, they are taken to a web page that encourages them to sign up for the product. This is a viral payoff page, and they have become a standard piece of the product marketing playbook in recent years.

The nature of team-based or multiplayer business software is that non-customers frequently come into contact with the tool as they work with its customers. This represents an opportunity for the product to gain new customers who just had an enjoyable experience using it.

However, most viral payoff pages are catching customers at a point where their need for your product might not be *highly qualified*. Because of this, viral payoff pages are often low-converting channels. For example, a product marketer may occasionally review website mockups sent over from their designer via Figma, but they don't have a need to create designs themselves, so they have no reason to sign up for Figma. There is certainly a place for payoff pages if your customer base is wide enough—virtually everyone needs a video meeting tool, for example—but recognize that it's no panacea.

Product teams can expect between 2–5% conversion rate on payoff pages, but must be careful about balancing sign-up intent vs. viral growth. If you make sign-up too easy, poorly qualified new customers may hurt your activation rates. While at Calendly, Oji spent an inordinate amount of time optimizing the company's viral payoff page to push toward that 5% threshold. The team focused on streamlining the sign-up process from the viral payoff page for new customers by pre-filling information from the scheduling process and reducing the number of steps required. This adjustment worked almost too well. Oji recognized that they had inadvertently made it too easy for low-intent customers to create accounts, which led to lower engagement and lower retention for some of these new accounts. By adding a bit of friction back into the process, the team lowered the viral sign-up rate down to roughly 3%. This filtered out the lowest-intent customers while still capitalizing on the viral growth potential.

Every synthetic viral hack has limits. Deploy them where the cost-benefit is high, but don't rely on them alone to build a viral product and business. The most viral products you know are paradigm-shifting products that continuously innovate to build a defensible position in the market while imbuing their brand with an aura of desirability and quality.

MEASURING K-FACTOR

We often get questions about how to measure virality. The simplest way is to measure the K-factor.

The K-factor is the average number of new users who have been invited to your app by your customers times the conversion rate of those invited users. (Average Invites × Conversion Rate). K-factor also has an element of a time period because customer invitations/referrals have to be timely to mean anything to your business.

In the real world, it can be hard to track invitations. Some happen over lunch or in communities in ways that cannot be tracked by your marketing team. So there is an even better way to measure K-factor that comes from the world of virology. In the simplest terms, if you start out with 1,000 customers in one period, how many do you end up with in the next period without applying marketing or factoring in churn? If you end up with 1,300 customers, then you have a K-factor of 0.3 or 30%.

Note that if your virality is greater than your churn rate, you can have a business that will grow without applying marketing dollars. Any marketing you apply should in theory just increase your growth rate.

NETWORK EFFECTS

Hopefully your sense of durable virality is rock solid. Let's turn to network effects. If durable virality is rocket fuel, and synthetic virality are afterburners, then network effects is how you reach orbital velocity—once you reach critical mass and create network effects, your product can draw in customers and create a self-sustaining ecosystem.

It's possible to achieve network effects without being viral; for example, telecom systems and financial systems have natural network effects. However, for most technology products, virality speeds up the impact of the deliberate design of network effects.

Network effects occur when the value of a network increases exponentially with every new node added to it. The classic example of a network effect is the telephone system. A single telephone is useless. Two telephones are marginally useful (i.e., only to the two people connected). A million telephones are world-changing because the potential connections to be made are astronomical. Today, there are twice as many mobile devices as there are people. The network effects of our communication system have taken over.

Here are more examples of network effects in action:

When a new customer adopts Google Docs, the number of documents every other user can potentially interact with and edit becomes exponentially larger. Google Docs becomes more valuable with every new user that adopts it.

When a new user joins Twitter, TikTok, or WhatsApp, the number of people every other user can potentially contact grows, along with their overall enjoyment of the app. In the case of Twitter, a user could suddenly start to see content from a world-renowned scientist or a journalist on the ground in a war zone. With every new user that joins the network, previous users are rewarded with a more valuable system. On the other hand, network effects are why it's so difficult to get a new social media platform off the ground. Users will always opt for the largest network, but to build a large network you need new users. It is a chicken vs. egg problem that few companies have solved.

Network effects are not just a property of social networks or communication channels. They can be engineered into most products which is why we are covering it in this chapter. It goes beyond growth to creating a sustainable advantage within that growth. At Calendly, Oji's team made it easier to book time with other Calendly customers. No more hunting about in your calendar to see times to match. The app shows you an overlay of your and your colleague's availability right in the interface. As more people join Calendly, each user's effort in scheduling is reduced.

Network effects are all around you. The phone network. The internet. Standards like 5G. Google Search (the more pages that are indexed, the better search gets, at least it used to be). Even the metric system (of course one stubborn country continues to not be drawn into the gravitational vortex . . .). When a product or business achieves network effects, it can become very large and scalable. Network effects are a powerful moat, making a business more resilient and more resistant to disruption. This is because customers tend to stay within that network as long as it offers continuous value, part of which increases every day as new customers are acquired.

There are two primary types of network effects in the software business:

1. **Direct network effects:** Direct network effects occur when the value of a software product or service increases for each individual user as more users join the network. For example, social media platforms like Facebook or messaging apps like WhatsApp become more valuable as more people join because the value of these platforms lies in the ability to connect and interact with other users.

2. **Indirect network effects:** Indirect network effects occur when the value of a software product or service increases for each customer as more complementary products or services are available on the same network. For instance, the value of a video game console increases as more game developers create and release games for that specific platform. More games attract more users, and more users attract more game developers, creating a positive feedback loop. You can also see the impact on large online marketplaces like Amazon and Alibaba, which have gained indirect network effects.

When Ezinne joined Procore, a leading construction management software platform, the company had identified an untapped network. While Procore had strong adoption among general contractors (GCs), it had much lower penetration among specialty contractors (SCs) like electricians, plumbers, and roofers. The product team spent months interviewing specialty contractors and discovered that they spent enormous energy discovering and landing new projects. This inspired Ezinne's team to build the Procore Construction Network (PCN), a platform designed to connect general contractors with specialty contractors. The PCN allowed SCs to create profiles showcasing their expertise and experience. General contractors could then search for and connect with the right specialty contractors for their projects. The network effects of the PCN had an immediate impact. As more specialty contractors joined, the platform became more valuable to general contractors looking for niche expertise—and as more GCs adopted the PCN to find partners, it became an indispensable lead generation tool for SCs. By connecting these two interdependent groups, Procore was able to create a virtuous cycle of adoption and engagement. The PCN has become a key competitive advantage for Procore and its customers.

Network effects are typically associated with consumer-facing technology, but they are just as relevant in B2B and enterprise. Too many B2B SaaS businesses stop at virality and don't push into network effects, and as such are more vulnerable to disruption and to slower growth. Product managers have to think beyond virality toward harnessing the power of network effects and standards to power their differentiation in the market.

There are a few things to keep in mind for engineering network effects.

Network Effects Require Critical Mass

You should not invest in network effects too early unless you are working on a standard or a communications network. Procore could not have built the PCN without a strong product suite and large user base of general contractors. Focus most of your early efforts on durable virality, but start to build the planks of a network effects strategy and key product investments over time. But keep network effects in the frame from the beginning—product managers are at their best when they can see the end from the beginning.

Pay Attention to Identity Systems

As we discussed in chapter 4, customer identifiers is a dangerously overlooked aspect of product management—and also a lucrative one. This is why companies like Google, Microsoft, and Apple all made sprawling investments in identity schemes. Each company wants to be your go-to identifier, which is why you can log in to so many apps using your Gmail or Outlook email. Apple came late to the party but has bootstrapped Apple ID on the strength of iOS. By making every app accessible via Apple ID, it makes it more likely that developers will build support for Apple ID into their apps, which in turn will increase lock-in for the iOS ecosystem—resulting in network effects.

Every successful software company will eventually have multiple products. Design your identity system to scale to many products as well as potentially external products via a developer program. You may never get to that scale but you will be glad you don't have to re-engineer everything if you do.

Build Features that Enhance Being a Member of the Network

As your product grows and your customer base expands, you can create additional value for your customers by building features that specifically enhance the value of being part of your product's network. This is often how network effects get bootstrapped if you're not a communications tool. It accelerates the critical mass needed to have a mature network effect.

The Procore Contractor Network is a perfect example: Both general contractors and specialty contractors benefit from being part of the network, and the value grows with each new contractor.

Calendly's availability comparison feature is another example. If you're scheduling a meeting with another Calendly user, Calendly overlays the person's options with your own availability and shows a green dot next to the times where you're both available. This saves Calendly customers a few precious seconds from having to open up their calendars and cross-referencing with the other person's availability. Small benefits like this add up over time.

Developer Ecosystems Can Create Strong Network Effects

When a platform provides developers with the tools and resources they need to build and distribute their own applications, it can create a self-reinforcing cycle of growth and value creation.

Apple's iOS platform has benefited from this for almost two decades. By providing developers with a robust set of APIs and a centralized App Store for distribution, Apple was able to attract a massive community of developers to its platform. As more developers built apps for iOS, the platform became more valuable for end users, who benefited from a wider selection of high-quality applications. This, in turn, attracted even more developers to the platform, creating a virtuous cycle of growth and innovation.

Compare iOS's massive app store compared to, say, ChromeOS, and you see the enormous impact of developer-driven network effects.

A recent example, as of this writing, is OpenAI's GPT marketplace that allows anyone to create and sell custom GPTs built on OpenAI's platform. There will be many developer platform plays like this as the AI supercycle kicks off. It's almost guaranteed that the winners will have compelling developer platforms as part of their core value propositions.

Define Industry Standards

Defining industry standards can be a powerful way for companies to create and capitalize on network effects. By establishing a dominant standard, a company can create a virtuous cycle of adoption and value creation that makes it difficult for competitors to catch up.

Tesla's Supercharger network perfectly illustrates the power of this strategy. By investing heavily in a proprietary network of high-speed charging stations and then opening up the technology license free to other car manufacturers, Tesla effectively forced the industry to capitulate to its standard.

As Tesla's market share grew, the Supercharger network became increasingly valuable, making Tesla's cars even more appealing to consumers. Other car manufacturers were left with a difficult choice: either invest in their own charging networks and risk fragmenting the market, or adopt Tesla's standard and help to further entrench the company's dominance.

By defining the industry standard for electric vehicle charging, Tesla was able to create a powerful network effect that not only fueled its own growth but also made it incredibly difficult for competitors to challenge its position.

Network effects are the holy grail of the product-led model. Howeverb its usually a culmination of many careful steps in the journey. Network effects are impossible without first identifying a sharp problem, listening to your customers, designing an elegant solution, onboarding and activating customers, pricing appropriately, measuring the right metrics, and in some appropriate cases, building virality. Every piece of the product-led puzzle is interconnected and dependent on the others. We hope we made that point clear in Stage 1 of this book.

In Stage 2, we will turn our attention from building PLG products to building product-led organizations.

STAGE 2

The Shipyard: Building Product Systems and Teams

The shape of product management leadership is changing.

Product leaders at all levels have to orchestrate far more than their titles imply, making sure the entire company—even beyond the product team—works well together and is consistently performing at a high level. The typical product-building organization has traditionally included engineering and design, with product managers tying them together in a triad. In a product-led organization, though, this triad is expanded to include data analysis, research, product marketing, and even customer support, customer success, and sales. The entire customer experience must be accounted for and made excellent, and it's often the product leader's job to pull it all together.

We have spent a collective forty-five–plus years in product management roles. We wrote this section for all technology company executives: founders, CEOs, CROs, and of course, Chief Product Officers. But this section is also for product managers who aspire to reach these levels in their career. It is never too early to start learning how world-class product organizations are run.

The first priority for product leaders is to ensure *the right talent is hired, nurtured, and trained to do good work*. Next, their job is to build and manage cross-functional product teams that tie together proirities from across the company. We call this expanded product organization the *Shipyard*. The product leader's role is to ensure everyone in the Shipyard works together seamlessly to create the best possible business outcomes.

Here are some of the other concerns and responsibilities product leaders:

- Providing a clear sense of direction, a clearly articulated set of features and product investments that will help the product win more customers

- Cultivating a healthy and motivated culture within the Shipyard that has strong doses of growth mindset and continuous learning

- Ensuring the entire company understands how to generate customer insights, and to harness and examine customer data in order to improve customer outcomes

- Partnering with all other business functions to optimize the customer experience beyond the code and the product to deliver on better business outcomes

Finally, very senior product leaders also materially shape the direction of the company, often working with the CEO, CFO, and COO to craft the company's strategy and clarifying how its product strategy enables it. The Chief Product Officer (CPO), with the help of the Chief Technology Officer (CTO), ensures harmony between the various levels of value: code, product, and business, at the highest levels of the company.

The shift to being product-led is often championed by a company's product leader, but it is not their job alone. While product leaders are becoming more prominent, the responsibility of building the best possible product (and therefore, business) is shared across boundaries into sales, engineering, and marketing.

Because of this shift, it is imperative that product leaders think beyond product building and think about being product-led as a company-wide directive.

Product Leader Tools

For a fleet of templates, worksheets, and playbooks for product leaders—designed and used by Oji and Ezinne—get the Pro Edition of *Building Rocketships* at productmind.co/brpro.

NINE

Product Systems: Building Strong Cross-Functional Culture and Teams

In the 1990s Microsoft decided to get into the personal finance software business after a failed attempt to buy Intuit in 1994.[19] This was a grave threat to Intuit and its core product, QuickBooks. Microsoft had a history of entering niches, like office productivity, and using its existing market power to drive out incumbents. Many analysts left Intuit for dead, but they didn't just roll over.[20] Because of their hyper-focus on personal finance, Intuit was able to consistently out-innovate and out-ship Microsoft, despite the giant's deep talent pool and deeper pockets. Microsoft eventually exited the market after losing to Intuit for many years. Today Intuit is still at the apex of personal and small business financial software.

How did Intuit defeat the largest software company on Earth? Oji was at Microsoft at the time and was close friends with the product manager who led Microsoft Money. With a front-row seat to this epic battle, it was clear that Intuit

---

19 "Buying Intuit," n.d. https://cs.stanford.edu/people/eroberts/cs181/ projects/corporate-monopolies/dangers_quicken.html.

20 Fried, Ina. "How Intuit Managed to Hold off Microsoft." *CNET*, June 11, 2009. https://www.cnet.com/culture/how-intuit-managed-to-hold-off-microsoft/.

outperformed Microsoft due to its superior *product system* that drove continuous, customer-focused innovation and rapid iteration.

Product systems are the practices, processes, principles, tools, and rituals that a technology company puts in place to ensure the making of high-quality technology products that meet their business and customer goals.

Product systems must be designed with intention. In order to keep pace with change, scale, people, and culture, product systems need built-in feedback loops to refine a company's approach to solving customer problems over time. Intuit's product system wasn't just confined to product management—it incorporated engineering, design, and go-to-market, all of which worked hand in hand to deliver better products that customers loved. Despite being pivotal in defining the modern product manager role, Microsoft was slow to innovate on effective customer-focused product systems.

Product systems help us continuously answer three questions:

    a.   What are the sharpest customer problems that will lead us to accomplishing our business goals?

    b.   What is the best solution to build to deliver on the customer problem and opportunities?

    c.   What is the best sequence to deliver in for maximum customer and business impact?

(Remember that in chapter 1's discourse on *roadmaps and sharp problems* we asserted that good roadmaps are basically building solutions in sharp problem order.)

In early-stage startups, the company itself *is* the product system, with virtually everyone working together to answer the three questions above. But as a company grows—especially if it takes off like a rocketship—the company's product "system" morphs into a Frankenstein's monster of informal and inefficient practices.

Every business has a product system, but most are not *intentionally designed*. The component parts just seem to happen—squads, Scrum, meetings, PM frameworks, tools, etc.—and they all operate independently from each other. The skeleton of the "system" is typically set up by the company's first product leader, and future leaders and teams inherit it, often not sufficiently questioning if or how it should evolve. They may tinker on the edges (say reducing sprints from two weeks to one week), but the core of the system calcifies. It becomes "just the way we do things" and slows innovation as the company grows.

We've seen many informal product systems in our careers. As vice president of an incubation team at Time, Inc., Ezinne initially struggled to define how to build within a scrappy innovation group. There was no consistent process for prioritizing ideas, validating them, and incorporating them into the larger business. This made every attempt to innovate a bespoke challenge. Oji, as head of product for Atlassian's communication group, inherited a system that spanned teams across continents, from Silicon Valley to Sydney and beyond. Very few things about how the teams functioned and coordinated with each other was codified. This led to inconsistent practices, delays, and frustration.

In both cases, the product organizations grew informally as the companies evolved. Both Oji and Ezinne spent countless hours redefining basic frameworks for product teams working together: conducting customer discovery, writing product briefs, outlining dependencies, setting communication rituals, creating roadmaps, and more.

In our experience, high-performing technology product teams are successful less than 50% of the time—meaning their product improvements and new features don't have the intended business impact. Most product teams come nowhere close to that level of shipping meaningful products. For ambitious folks on inefficient teams, it can feel like running in place and wasting precious time on BS work. Most of us get into technology to solve hard problems and make a corner of the world a better place, but too often our creativity is stifled by chaotic systems (e.g., disjoint or missing processes, proliferation of direction-setting artifacts, unnecessary rituals and meetings). How many of us have been on teams where we have shipped things that don't matter? How many hours have we spent building software features that will go into a black hole of lost productivity for your team and almost no consequence for your customers?

This is especially true in large companies whose primary business is not selling technology, but for whom information technology is a large input[21] to return on investment. This is also heightened in regulated industries. Our longtime collaborator and consulting partner, Ted Yang, has done multiple tours of duty as a technology leader at investment banks:

> Banks derive significant edge from internal software development and so [they] spend a lot of time building tools. They often have teams who report into a parallel org structure from the businesses that use those tools. Without exception, banks have complicated product processes that were driven more by control and risk management than they were by any desire to get things done.
>
> In fact, senior technology leaders would privately admit that one of the purposes of existing processes was to slow down the stream of requests from business customers. And they would admit that many of the component parts (especially documentation) served little to no purpose—they were "the way it has always been done." Any attempt to streamline or revise processes would be swamped by more urgent concerns or by political objections from those who like things the way they are.

When companies invest in features that customers don't particularly care about, they waste and mismanage vital human capital. We never tire of reminding PMs that as leaders, they are often responsible for millions of dollars in *people resources*. But this is not just business waste, this is also a wasting of the human spirit, our ingenuity, our life force. Meaningless work can slowly kill the drive in a team and every useless feature we ship is another cut.

---

21   Software is eating the world, so every company is now a software company in some sense.

This is why developing intentional, well-designed product systems is so important. *We're attempting to harness the human ingenuity we have to the best possible extent.*

THE SHIPYARD: A COLLABORATIVE MODEL FOR PRODUCT-LED PRODUCT TEAMS

Product systems are *systems of systems*, making them somewhat ephemeral and invisible to many people in a company. So to really *see* the product system, we need to break it down into its component parts.

Product systems are not just for the product management discipline alone. It is intended for product managers and their multidisciplinary peers—the entire range of creative collaborators they need to ship useful products that move the needle for businesses. PMs can't accomplish much without their multidisciplinary peers.

There are many ways to describe a product system, but we use this three-tiered model for its simplicity and flexibility.

1. **People & Management Systems**—how to deliberately get the most out of the talent you have in the product organization, and given the context of this book, especially PM talent.

2. **Direction Setting and Strategy Systems**—how to choose what to go after so you can focus a product organization and deliver on growth.

3. **Execution Systems**—how to repeatedly build software that customers love and the business celebrates, even as your company grows.

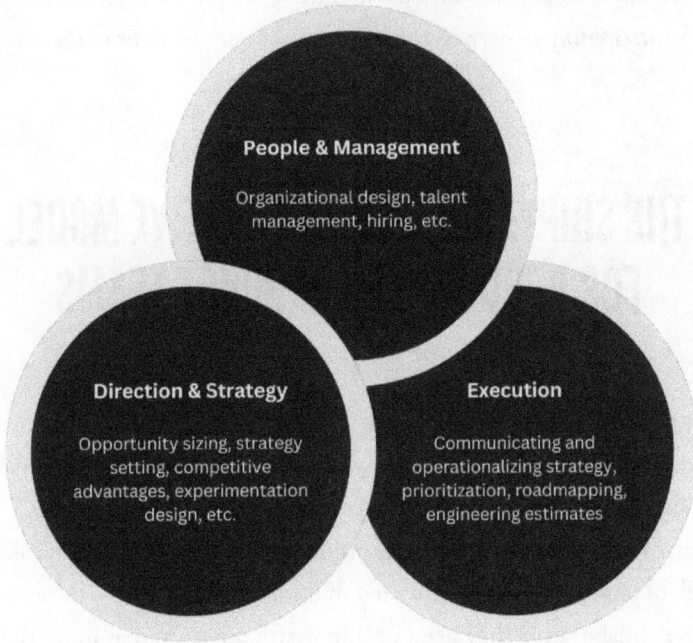

You might describe your product system in different terms, and that's ok. The grouping is not the most important thing, the *completeness and interaction* is.

We dedicate entire chapters to each of these three systems, so we won't dive deep into them just yet. For now, let's discuss how to keep the component parts working in harmony. Product-led product management teams are tasked with building rocketships—it's only fitting to imagine our product system as the Shipyard where that rocketship is built.

The Shipyard is a model for building product systems that are collaborative and multidisciplinary. It's built on a fundamental building block often referred to as a *squad*—a small, multidisciplinary team that can build and ship *almost* autonomously. Squads are given the responsibility of achieving key product and business goals, which they accomplish by executing on key projects through a prioritized roadmap. The larger and more complex the project, the more squads that will be assigned to work on it. Multiple squads can be tasked to go after big goals.

Squads have been around since the dawn of software product management in the 1980s, but the Shipyard reimagines them. Squads were originally composed of the EPD triad: engineers, product managers, and designers. EPD squads exemplified the value of multi-skilled, self-sustaining teams that could execute a single mission. Their key innovation was decreased product cycle time, or the freedom to see a project from start to finish before moving on to the next one. By focusing on one important thing, an EPD squad could collaborate more rapidly to solve creative problems and deliver on related business goals. Each squad member's focus meant a faster cycle time. Dedicated EPD squads worked so well that almost every advanced technology company has adopted them, and many business teams across the economy that use technology as inputs are following suit.

In theory, squads could work quickly and independently to deliver products and features. In practice, EPD squads had two limitations that hurt their productivity:

1. EPD squads were never fully autonomous. The EPD triad still relied on outside departments and resources to ship products: research, data analysis, product marketing, and more. These were often organized as service organizations, and these unaligned resources could become cycle time bottlenecks.

2. EPD squads were often beholden to their departments, not their project goals. If the head of engineering wanted to do a security review, engineers had to prioritize that over their squad duties. PMs were also often not dedicated to a single EPD; they would write specifications, hand them off to the designers and engineers, and move on to the next problem to solve. Sometimes, no one fully took responsibility for the final solution, and when they did, it was a major hassle to align effort productively, resulting in products and features that missed the mark. EPD squads were often unfocused, and the results showed.

The Shipyard model removes these limitations to fully unleash the power of squads. The first change is expanding the EPD triad into at least a six-part team:

- Engineering: to build the solution with tech

- Product Management: to coordinate and ensure we are working on the most impactful thing

- Design: to ensure the most usable solution

- Customer Research: to ensure the customer feedback loop is stood up and maintained

- Data Analysis: to ensure the learning loop is stood up and maintained

- Product Marketing: to ensure the story of the problem is clear across the company and to the market

Each Shipyard squad ideally has dedicated resources from each of these domains. These domains represent the full gamut of skill sets needed to traverse customer opportunity finding, right solution building, and market introduction. This shift increases the expanded squad's creative and execution capabilities and autonomy. Some examples:

- Product marketing can be fully involved from the beginning of a product- or feature-planning cycle, helping craft more nuanced product and launch stories.

- Engineers can be deeply embedded in the customer research process so the quality of their technology tradeoff recommendations can improve.

- Data analysis from data analysts informs every function within the Shipyard squad and helps the team make better decisions and iterate more quickly.

The typical ratio for a Shipyard squad—such as those run at companies like Microsoft, Slack, Meta, and Google—is five to seven engineers, one PM, one designer, and one-third of a product marketer, customer researcher, and data analyst each. (Product marketing, research, and data team members should have enough bandwidth to work in multiple squads.)

The Shipyard squad must also have close ties to three other departments: customer support, customer success, and sales. Squads need a constant feed of customer insight data from these teams to address customer problems and requests quickly. Communication is a two-way street; Shipyard squads can also help address more technical customer support and sales questions needed by customer teams.

Here is the critical feature of the Shipyard model: Local priorities are no longer set by departments, but rather by squad units. Shipyard squads need to work as a single unit that balances their team objectives with departmental responsibilities. Each department usually has a tax that means product marketers or PMs are answerable to their department heads. However, the Shipyard squad is encouraged to prioritize their team goal higher than the department's in most cases. Unless there is an emergency, Shipyard squads should be left to focus on their top priorities, as long as they are aligned with product and company objectives.

WHAT IF YOU DON'T HAVE A FULL SHIPYARD SQUAD?

The six functions of the Shipyard squad are almost always present as key tasks in good product-led companies, even if they can't yet afford a dedicated professional. For example, even if you don't have data scientists, either your engineers, product managers, or product marketers are (or should be) crunching some kind of data to make forward-looking roadmap decisions.

Think of each of these key functions as responsibilities for which we hire dedicated trained professionals to carry out, e.g., engineering is a core responsibility of a tech startup for which it hires engineers. Just like implicit product systems that people often don't notice, these responsibilities are always there, just sometimes being carried out by a non-expert until your company can afford to hire a dedicated person.

In small startups, the first few employees wear many hats. This is just a euphemism for one person performing many of these responsibilities at once. For example, the engineers also write tests and do data analysis as needed.

Not having dedicated experts for all six functions is not a problem. Just be sure all squad members are committed to wearing many hats to solve product and business problems together.

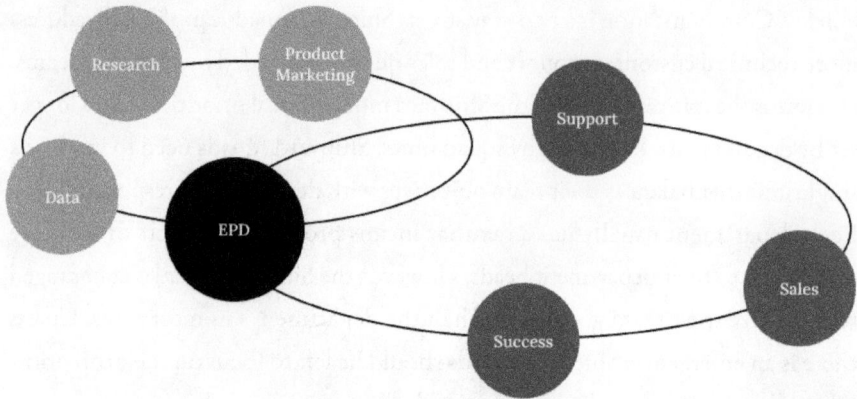

The Shipyard squad expands on the principles of the original EPD squad: focused, multi-skilled teams that have a singular mission. Specifically, Shipyard squads create two key advantages:

- **They can move faster.** This is because of team familiarity, clear goals, quicker decisions and mandates, and ability to go from idea to market quickly. If you hold complexity constant, some squads can deliver at 2x the speed of traditional EPD teams when counting the time it takes to get products in the hands of actual customers.

- **They can handle more complex projects.** Traditional discipline-focused teams can often get bogged down because managers need to absorb fine levels of detail from the team to help manage complexity vs. having the team use domain familiarity to solve complexity by itself. As the technology industry contemplates more and more ambitious projects, we need better ways to manage complicated projects and still deliver usable and lovable products.

Shipyard squads improve collaboration, reduce cycle time, and most importantly, can often deliver end-to-end product experiences that delight customers and maximize outcomes.

GOOD SHIPYARDS NEED COORDINATED LEADERSHIP

The product system is really *the operating system for the humans and management that make the product*.

The Chief Product Officer, the Chief Technology Officer, the Chief Design Officer, and the leader of product marketing—these are the usual ultimate leaders of all the functions of the six-part squad—must create a cohesive and adaptive *innovation machine* to build what is needed to win their competitive market.

They have to persuade the cooperation of the rest of the organization in this endeavor, especially the leadership of the non-maker customer-facing teams.

The anti-pattern we often see is that there is no effort to find a shared way of working for these creative professional disciplines, leading to execution inefficiencies. It doesn't really work when Shipyard squads are in conflict with other departments that don't really understand or resist how they work. True transformation comes from a lot of leadership harmony on how things are done together.

TRANSFORMING YOUR PRODUCT SYSTEM

Change management of product systems is hard. It can be difficult to change a system that is used by tens, hundreds, or thousands of people. It can be hard to justify the time and effort to re-teach and re-train. It can take time to foist change on product management, engineering, and design leaders, who are wary of lost productivity, especially when the end result can be hard to quantify immediately. And in most cases, switching costs are magnified, making many people wary of them.

In practice, many change efforts fail due to tissue rejection—the existing organization cannot absorb it. A straightforward phased implementation of a new, better system may have lots of initial buy-in, such that implementation seems obvious. But it can ignore the fact that most people are motivated by just getting their current day job done in the way it always has been, versus an uncertain redesign that may hypothetically threaten the stability of their here and now. Let's be real—many professionals are burnt out from constant organizational change.

Gradual change is also not a certain panacea. Often, existing systems reject any attempts to change in a piecemeal fashion. What's worse is that many organizations will plan change management for product processes, but then declare victory too early so their teams can "get back to work." In those situations the new processes tend to revert due to inertia, frustrating many.

So how do you go about building, updating, or transforming your product system? We share a few concepts that have worked for us next.

Identify Gaps and Quantify Risks

Most product leaders, from PMs to Chief Product Officers, will perform an evaluation of all their product subsystems when they start a new job. This is done either implicitly or explicitly, though the latter is better. We advocate for creating a product systems checklist. We share our Shipyard System Checklist in the Pro Edition of this book.

Map your evaluation to the product system simplifying and its component parts. This will do two things:

1. Help you identify the risks of change
2. Give you a prioritized list of what needs to change.

It should be no surprise that the core of a product leader's work is about prioritizing what systems should endure and which should change.

Problem Acknowledgment and Buy-In

When you have identified systems that need adjustment or wholesale change, you need to socialize your identification and need for change. Changing a product system is more fragile if there is no real sense of acknowledgment by most people in the organization about the need for change. Product leaders cannot act on the conclusions of their evaluation in isolation. They need to confirm their assessment of the risk and priority of the gaps in the product system socially, and then telegraph their intention to act to change it.

Members of the Shipyard, when confronted with the need for change, need to have a fundamental sense that the problem exists and the change is needed to improve everyone's performance.

If this doesn't happen, leaders will lose credibility and will be seen as spending time on pet projects that will have no impact.

Get High-Level Support

Most change, when enacted, will require effort and likely push out other existing priorities. It's important to get buy-in at the highest levels of your organization. For product leaders, this typically means the CEO and the rest of the C-suite. Share your evaluation, risk assessment, and the intended result of change, including its benefits. Often the other leaders are impacted and they are usually very interested in whether that impact is positive or negative. You should also convince your peers that this change is worthwhile, even if they don't directly have a hand in the transformation.

Get Grassroots Support

Executive support for product system changes always gives you some leeway. Now turn around and get the same from your management team, Shipyard leaders, and product managers. This makes the difference between reluctant change makers and enthusiastic drivers of change.

We recommend working hard to get buy-in as low in the organization as possible. We talk more about this in chapter 10 (Harnessing Talent), but there is definitely room for populist tools of leadership in order to reconfigure how the Shipyard works effectively.

Change What Matters

Some aspects of your product system may be just fine! Maybe thoughtful predecessors designed it. If so, just thank the ancestors and be grateful, while making that part of the product system work for you.

In other cases, some parts of the product system are just not worth fiddling with at the moment, based on your assessment.

Tearing down the product system and starting over can be tempting. But organizations can get change fatigue when it goes on for too long. Think about what matters, what improvements will make the highest impact, solve multiple problems, and do those first.

Communicate Benefits Constantly

After putting in effort, Shipyard professionals and teams want to know if they are better off from the changes we asked them to make. They need a feedback loop.

Shipyard design and change should come with a corresponding commitment to communicate how those changes have worked or not worked. And if they have, a quantification of the improvement is always very helpful. Share the change in shipping speed, customer satisfaction, and product adoption. If the transformation does *not* have the intended effect, address it head-on and rally your team to find a new solution.

Hire a Caretaker (Product Ops)

This is an overlooked key to success for designing and changing your product system: Get a dedicated product system caretaker.

These professionals are now often referred to as product operations and they are truly one of the most critical roles in a modern product system. Refining and maintaining the product system in decent-sized organizations is a full-time job. The operating system needs constant tuning, because it's full of the most volatile element in organizations: people. Just like agile development systems have *Scrum Masters*, Shipyards need *Product Ops*.

A good product operations professional can fully focus on making squads work—establishing their working templates, tools, cadence, and rituals. They observe, reinforce, upgrade, update, troubleshoot, and tune. They become an essential advocate other than the chief product officer for everyone in the Shipyard to solve operational and innovation problems. One specific failure mode we have seen over time is when CPOs spend all their time manning product operations, only to neglect other key responsibilities they have.

Don't Make "System" a Dirty Word

Next-level product organizations always have people who innovate faster and better than the product system caters for. They tend to find ways to productively break the base process to do better and ship faster. This is a good thing. *The system is always in service of and subservient to faster and smarter innovation, and not the other way around.*

Therefore, it's ok to make exceptions (i.e., allow the innovators to move faster) as long as the team hews to the essential spirit of the system. Faster routes should never simply cut corners or go against customer sense, without some other well-articulated and outsized benefit. Smart leaders (and their product ops) try to understand, adopt, and document these *system exceptions* when they work well, and then try to quickly graft them into the established process. Once Ezinne started to socialize the changes to the product systems she was going to make at Procore, one of the incubation teams pointed out where they need exceptions. She brought them in to co-design a system that would work for them, but also for non-incubation teams.

In summary, the most important ideas in creating product systems are:

1. They are necessary and need deliberate design.
2. They must be maintained to stave off entropy.
3. They must be malleable and changeable or they will not serve the organization very long.

Product systems are never complete; they are in a constant state of adaptation and improvement. Product systems are the floor of the process of innovation, not the ceiling. They help to make innovation repeatable, which is necessary to win consistently in your chosen market.

In the next few chapters, we will dive deep into the three component parts of The Shipyard product system: people systems, strategy systems, and execution systems.

TEN

Harnessing Talent:
Systems of People and Management

People who find themselves in Shipyard organizations at software companies are ambitious. They want career growth, purpose, autonomy, and mastery—usually in that order.

Product managers in particular are outliers on this spectrum because the profession attracts people with outsized ambitions, A-type personalities, and aspiring entrepreneurs. It seems like every other PM we have met in our careers wants to go on to be a founder and/or CEO. This kind of ambition is a good thing when they are prepared to work hard, therefore the most important job of the product leader is to keep these kinds of people motivated, productive, and happy.

The value of a good people system is *effective leverage of talent*, which is by far your most significant actual cost in a Shipyard organization. More importantly, it's the most significant driver of value. For systems of people and management, the key questions for product leadership at all levels are:

- How do you attract, assess, and improve talent within the organization?

- What approach and methods do you rely on to connect your people, their skills, and ambitions to the company's objectives?

- How are you designing your organizations and the shape of your teams so it gets to the heart of what your company needs?

Here are the key steps for building an effective people and management system in your product organization.

DESIGNING YOUR SHIPYARD

Team building is not just a people question, but a resource allocation and investment one. Once you have a clear product strategy, the next step is to decide what resources you will need—and where—to achieve your goals.

As we introduced in chapter 9, high-performance product-led teams are organized into cross-functional and semi-autonomous squads. Squads serve as a good proxy for measuring how much effort is being put into any specific part of the product strategy. Therefore, as we discuss each variable of team building in this section, remember that the final answer comes down to your priorities.

The New Squad

In leading-edge high-tech companies circa early 2000s, squads were composed of software engineers, product managers, and designers, which made up the EPD triad: A team might have 4–7 engineers alongside one PM and one product designer. But as speed and complexity have risen in the tech industry, it has become harder, if not impossible, for squads to work completely autonomously. The core EPD triad relied on too many people outside the squad to do good work, including data analysts, researchers, product marketers, and even sales and customer success. We needed more integrated squads to execute effectively.

The Shipyard Squad

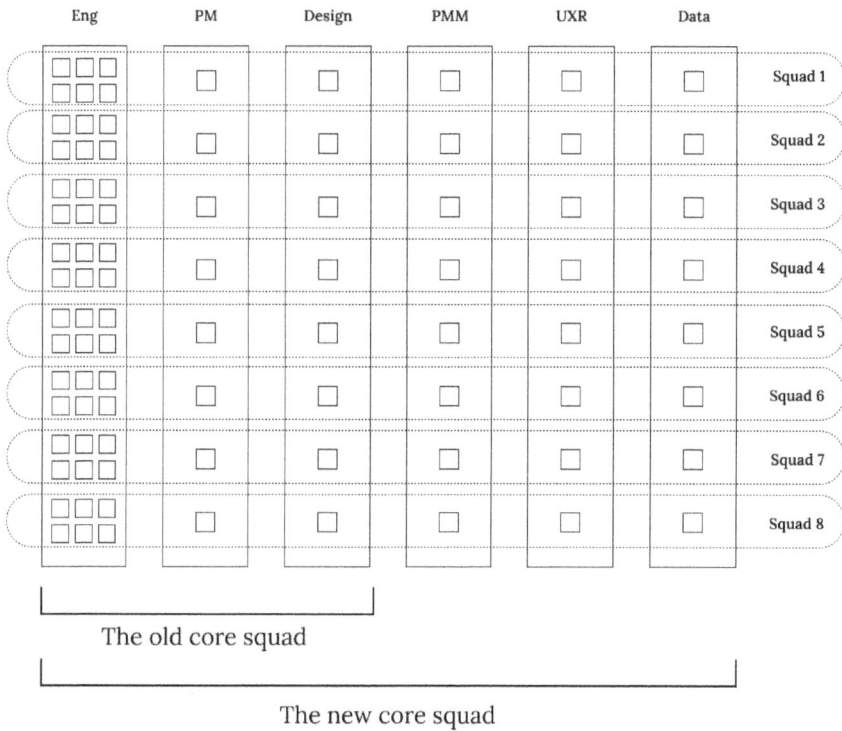

Eng	PM	Design	PMM	UXR	Data	
□□□ □□□	□	□	□	□	□	Squad 1
□□□ □□□	□	□	□	□	□	Squad 2
□□□ □□□	□	□	□	□	□	Squad 3
□□□ □□□	□	□	□	□	□	Squad 4
□□□ □□□	□	□	□	□	□	Squad 5
□□□ □□□	□	□	□	□	□	Squad 6
□□□ □□□	□	□	□	□	□	Squad 7
□□□ □□□	□	□	□	□	□	Squad 8

The old core squad

The new core squad

This need for tighter coordination is what inspired the Shipyard mentality of running product teams. Remember, we recommend a larger "core" squad with product marketing, research, and data analysis added to the team. Some of the squad roles typically don't require a full-time person; for example, we usually aim for about 3:1 or 2:1 time for the product manager to product marketing manager staffing ratio. Also remember that the squad closely coordinates with sales, customer success, and customer support. These roles spend a ton of time with customers and the Shipyard needs to integrate that intel effectively to continue to be customer-focused. Product leaders should coordinate bilateral meetings between squads and the key members of these departments to share insights and customer feedback. There may also be new rituals and artifacts developed to keep the core and extended teams coordinated.

Fully-Integrated Shipyard Organization

The opposite of the Shipyard mentality is the "relay race" mentality. Too many organizations treat product as a thing handed off from department to department in a serial fashion. The proverbial "throw it over the wall and be done with it" way of working will lead to half-baked products and countless delays. The Shipyard must work as a cohesive system. When it's working well, the Shipyard will feel like controlled chaos, with communication and ideas moving on well-traveled paths within squads, with departments seamlessly keeping abreast of both team performance and department goals.

Multidisciplinary teams share responsibility for outcomes set for their squads, meaning individuals receive both a team grade and an individual grade for their efforts. Both the team outcomes and individual performance matter during performance reviews. For example, usually, a PM's team delivery outcomes (i.e., features shipped and their impact) matter to their performance review as well as their personal expression and growth in their particular PM craft.

Team Size

Squads are the indivisible unit of productivity within Shipyard organizations. You will rarely allocate resources one individual at a time, but rather by autonomous, multidisciplinary teams. For example, a new feature to a core product may require one squad, while a new product may be comprised of three to ten squads, coordinating to build the new thing. This cohesive structure allows the entire product team to align more deeply with the customer voice and business objectives.

A squad should only be as big as needed to achieve an objective or solve a very specific customer problem in a discrete amount of time. They should be large enough to focus on one creative feature-sized problem and solve it in the shortest amount of time that can materially impact the business.

Bigger teams of ten to twelve can be deployed to work on multiple related problems or one large, complex problem. However, larger teams require coordination tasks and a skilled manager to stay on top of the work to prevent slowdowns. This is the main tradeoff of larger teams—more coordination effort.

Larger squads are not always better; more engineers can actually lead to reduced productivity as coordination takes precedence over building. Beware of larger configurations, unless you can effectively divide the work with few dependencies and little coordination overhead. Twelve members are the largest we recommend for a single squad (not including sales, support, and account management). Be very cautious of managers who try to articulate a need for larger teams without articulating how to deal with these scale problems effectively.

Instead of growing the size of the squad beyond its natural boundary, product leaders should assign multiple squads to a problem if its size, complexity, and importance warrants it. This is what we mean by resource allocation: Squads are not just teams of talent, but *units of work capacity*. The more complex the project, the more units (squad-sized resources) should be assigned to it.

ALWAYS BALANCE SQUAD TUNNEL VISION WITH OVERALL PRIORITIES

Squads and teams have a weird kind of inertia. They love to work on mission-related problems for as long as they can—this is not only what we usually ask them to do, but it's a sequence that allows them to develop deep expertise. This typically means they start with the biggest-impact problems within their mission scope and work their way down the list of priorities. However, if a team is good, pretty soon it will have made a lot of progress on the big problems and will start to work on low-impact problems in its problem space, mostly refining things, instead of blazing new trails or solving deeper problems.

In other words, squads can develop tunnel vision. When this happens, team and group managers must carefully redirect and assign teams to the biggest problems in their general problem space. It's ok to ask teams to work on new problems they may be unfamiliar with. In fact, set the expectation that *a new mission may be assigned every so often* so that it's an expected part of your culture.

When teams expect mission reassignment and can get excited about the prospect, it makes your organization more flexible and agile. You want everyone on your team focused on accomplishing the main strategic goal, not just their localized objectives.

Team Skills

Squad size is not the only consideration. Teams need the right mix of skills to deliver quickly and autonomously. This might take some trial and error.

At Atlassian, when Oji headed the communication product division, the team had to ship features across all consumer platforms (i.e., Web, iOS, and Android) fairly simultaneously. The squads were comprised of frontend, backend, and services engineers, but no mobile engineers, who were part of their own group. Once Oji's teams were done with the web app, they would throw it over the wall to the mobile group to create the iOS and Android apps. The mobile group was chronically understaffed and was not as intimately aware of the features they were tasked to build because they were generalists. This often led to late delivery and poor quality. After diagnosing the issue, the division's squads were reconfigured to include iOS and Android engineers. The new mobile team members became experts in the problem space they were working on, which led to fewer delays and much higher-quality work.

This was a specific optimization for Atlassian's communication department and was unnecessary in other divisions of the company. Every organization and every department will need a different mix of skills on their squads. Another example: while Oji was at Twitter, they almost exclusively valued mobile engineers since iOS and Android were by far their biggest platforms. Squads had a ratio of 2:1 mobile engineers to web and backend engineers combined—an entirely different configuration to web-first companies like Atlassian.

The size of your company is also a factor. While not a hard-and-fast rule, there are high-level patterns we can look at for inspiration. Here are some rough guidelines for skill allocation based on company scale:

- **Startups**—Work to build a good core product team *and then* a growth team. A product team works on the core value you deliver and innovates on it. A growth team works to improve acquisition and activation metrics once you achieve some measure of product-market fit.

- **Mid-scale companies**—Add additional core product squads that focus on different product lines or specific customer segments. Then create shared resource teams like a developer tools team (to make every developer more productive), platform team (to serve your developer customers and partner integrations), and security and site reliability (to keep the infrastructure running at high reliability and unhacked by bad actors).

- **Large, scaled companies**—Specialize a bit more to focus and add capacity. Add dedicated data science teams (to make sense of product analytics), larger and more capable customer research teams (to deeply understand the customers you have and the ones you want), and dedicated customer feedback organizations (to make sense of what millions of customers are saying about your products). Some large companies have specialty teams like a pure research arm, incubation teams, and heist teams that work on specific opportunities (zero to one) as they appear, etc.

There is no one-size-fits-all squad configuration that is right for every company. The proper mix of skills will change depending on the business and division. Again, the goal is to optimize for fast, high-quality, and autonomous delivery of new products and features, so the skill mix has to be considered carefully.

Pro Edition: Skills Mapping Template

It's important to understand the skills you currently have on your team and the gaps you need to fill. One technique for doing this is called skills mapping. Get a full explanation of skills mapping, plus our go-to template, in the Pro Edition of *Building Rocketships:* productmind.co/brpro.

Team DNA

Company cultures attract certain kinds of people who are good at building specific kinds of things. Let's call this concept "team DNA." One extreme example is a company like SpaceX, which attracts some of the world's best rocket and aerospace engineers. Software companies also tend to be magnets for certain types of builders.

Procore, for example, attracts many engineers, product managers, designers, and others who have an affinity for the construction industry. Many have worked in construction or real estate previously. Hiring for a certain DNA is not always intentional, but it can help accelerate a company's learning and development curve when building new products.

Not having the right DNA can be a real detriment in a competitive industry. When Microsoft launched the Xbox, it didn't have gaming DNA, which put it at a disadvantage compared to competitors like Sony and Nintendo that had been cultivating their gaming culture and talent for years. As a result, the first couple of generations did not have outstanding success. Even today, after retooling its teams from successive generations of Xboxes, Microsoft still struggles to fend off smaller players who are more adept at imagining next-generation products and consoles.

Most managers and product leaders won't think about team DNA until they have a new problem to solve. Then, it often becomes painfully clear whether or not the team has the right passion and domain expertise to tackle the problem. Team DNA has nothing to do with pure technical skill; it's about having the intuition and creativity needed to make good decisions, day in day out, that are adequately reflective of customer tastes in that specific problem space.

Managers who recognize their team's DNA deficiencies should try to hire creatives with the right stuff, but they also have to beware of placing these people within teams that don't appreciate the value of cultural knowledge, thus suffocating or marginalizing them. It often becomes necessary to create brand-new teams with the right DNA for a new domain rather than try to retrofit a team with the wrong DNA.

HIRING AND MANAGING TALENT

If we had one piece of advice for product leaders on hiring and managing talent, it's this: Design your organization and the exact roles you need first, *then* hire the talent needed to fill those roles.

This principle was ingrained into Oji during his time working at Ray Dalio's Bridgewater Associates, one of the largest and most successful hedge funds in the world. At Bridgewater, every role had to be precisely defined by hiring managers before it was filled either by internal or external candidates.

Each role had to map *specific skills, personal values, and attributes* needed to succeed. You had to paint a crystal-clear picture of who would be ideal for the job. It went much, much deeper than the typical list of job requirements. Here is an example of an internal job description we would write at Bridgewater:

We need a platform product manager who knows how to build REST APIs, who is scrupulously fastidious and honest. They also need to be a self-starter, with high motivation, high visualization, who takes the bull by the horns, with high conflict and low respect for authority. (Bridgewater's job as a hedge fund was to bet against consensus to find hidden opportunities in the market, so they needed people who were not just comfortable going against the grain, but reveled in it.)

As a result of their stringent standards, hiring at Bridgewater was often painfully slow, but their success is a testament to the practice. Precise hiring is by far the highest leverage action you can take as a product organization. But what does that look like in practice?

Hire Precisely

The most effective product organizations are incredibly organized and principled in their hiring.

Once you map the skills and values needed, they should be written and shared widely throughout the organization, and especially with hiring managers and recruiters.

However, in a scaled organization, there are various flavors of any discipline that may be more tailor-made for what you need, for example, platform PMs, security PMs, UX PMs, etc. Each of these roles—though similar—should be defined distinctly and hired for separately. Each of these will then have an interview plan—a clear, consistent definition of experience and skills needed, clear questions that assess those skills, case studies that go deeper (if your organization does those), and so on.

The most important thing is that your hiring plan is thought out ahead of time. An organization needs to clearly define the ideal candidate, the exact skills they will need to succeed, and have a clear plan for discovering that talent via job descriptions, pre-prepared interview questions, level calibration, and more.

Implement a Hiring Refinement Loop

Hiring is an art and probably always will be one. What we attempt to do in product systems, especially the people system, is to remove as much variability as possible and to be a bit more calculated. But since we are dealing with humans, it's impossible to make this a purely scientific process.

Thus the best defense is to learn quickly from failure by implementing an improvement feedback loop. There is a principle at play here—everything you do many times requires ongoing optimization to improve performance, especially if it's a non-deterministic process.

We recommend that periodically—every quarter, half year, or annually, depending on the organization—leaders review the hiring system. Examine where it failed and make adjustments and update the process. We believe continuous learning can be applied to this important engine for making the shipyard excellent.

UNDERSTANDING THE DISTINCTION BETWEEN SKILLS AND COMPETENCIES

Skills are specific, learned activities that vary in complexity, from simple tasks like mopping a floor or typing, to complex ones like performing brain surgery. They indicate what a person can do, based on their training and knowledge, and help determine their preparedness for specific workplace activities. Skills provide the "what"—the specific abilities needed for a job.

Competencies represent how well an individual leverages their hard and soft skills, and their motivations to perform successfully in a work environment. They encompass the observable behavior that demonstrates "how" a person will think and behave and the "why" behind those choices as they determine the "what" that drives success.

Over the years, we've come to subscribe to this simple formula:

COMPETENCIES = SKILLS + ATTRIBUTES + VALUES

Attributes are characteristics of a person that describe how they tend to feel, think, and behave, such as curiosity, adaptability, and being detail-oriented.

Values are fundamental ideas and beliefs that guide a person or organization's motivations and decisions, such as transparency, and being helpful.

Both skills and competencies reference abilities acquired through training and experience. However, their roles in talent management differ significantly.

In our opinion, skills, attributes, and values are typically listed in job descriptions and are easily tested and screened for in interviews.

Competency, however, often requires observation. This may be the reason for the rise in popularity of case studies—as hiring teams aimed to observe how candidates combined their skills, leveraged their attributes and leaned into their values to solve what they considered a "true-world problem."

Competency guides are also valuable after recruiting in the career development stage where you are helping employees grow and find success at the company.

Craft a Career and Competency Ladder

As the workforce skews younger, it's important to outline how career growth can occur, what is expected at each level, and how it's judged. Without this, you will have consistently unhappy members of the Shipyard. They will not be able to connect their daily striving to promotion and mastery. Most of these professionals need something to aspire to as soon as their boots hit the ground. A career ladder has to be detailed and prescriptive about titles, competencies needed, expectations, and key defining deliverables. It should form the context of career coaching, management, and promotion conversations to be effective. (These things may not matter as much in a startup, because the adrenaline rush of making a company viable is enough for most people. But once there is product-market fit and growth is the main question, then this matters a lot.)

A career ladder is different from a competency ladder. A career ladder rationalizes the levels of authority and seniority in a company. A competency ladder attempts to articulate the increasing levels of demonstrated skills, influence, and outcomes needed to show competence at a particular level. Together, these tools provide a map to increasing career progression and competence. This is useful because it connects increased experience, learning, and outcomes to career progression—meaning that incentives are generally aligned toward more expertise in your company.

Recognize Success with Incentives

In order to stay accountable to pre-established goals, and because we are preconditioned by our educational system, the workplace needs a hierarchy of acknowledgments, incentives, and goal assessments. These can be things such as peer acknowledgment, monetary awards, or promotions. The most important point is to make sure that this overlapping system is designed, together with HR, to fit the desired culture of the workplace—to incentivize the attainment of minor and major goals, to acknowledge those who exhibit good company and team values consistent with growth.

Acknowledgments, no matter how small, keep a well-oiled workplace working well and hook into the natural dopamine cognitive systems of reward. It gamifies good work. From experience, we know that small rewards like mentioning a person in a Slack channel to indicate they did something amazing is a really big deal, so you should design a way to do this at scale.

As you verge onto the bigger, less-social recognitions, like monetary reward, work to make them transparent and based on best practices so you can use your resources well for more costly incentives like promotions.

Mentorship and Coaching

Coaching is natural in any human system—it's the way we work. But making it a core value can elevate your organization to a whole other level. For example, explicitly making it a core part of your organization's values by rewarding it informally but also at the level of promotions. The more senior a person, the more they have to share with their colleagues at lower levels, and they should be encouraged and compensated accordingly.

A mentoring culture improves outcomes for everyone, helps the organization adapt faster to changes in any discipline, and keeps unforced errors low.

ADVICE FOR PRODUCT MANAGERS WHO ARE EARLIER ON IN THEIR CAREER

If you're earlier on in your career, product systems can seem like they are above your pay grade. Nothing could be further from the truth. One hack in our careers that we used is working to be involved in crafting product systems. This requires extra time from building features, but it's a useful way to gain experience, gain the attention of leadership, and stretch your PM soft skills.

For systems of people & management, here are some tips on how to get involved no matter your level:

- Get interview trained and become practiced in hiring new talent. Be very thoughtful about hiring feedback and create a reputation for this thoughtfulness.
- Seek mentorship and make clear you want to be a mentor as a stepping stone to a future ambition to manage. This is generally a conversation with your manager and skip-level.
- Be liberal with kudos for your teammates. Become part of the culture of rewards and incentives. We have noticed for some reason that PMs don't do as many kudos as engineers or designers. Break the mold.

At every level, you can be involved in helping shape good product systems. Just ask to be involved and often, because of the toil involved, PM leaders will let you.

Hiring Mistakes

Don't be tempted into cutting corners in Shipyard hires. Establish a process and stick to it. Check to see if it's working well based on real-world performance and then tweak as needed.

The cost of making a bad hire far eclipses the effort required in the hiring process. The smaller the company, the more mission-critical each hire becomes. Especially in the case of product management, because it represents coordination and customer-focused headcount, which is *initially* harder to justify than engineering hires.

Bad hires have to be first identified as not consistently doing good work. Sometimes this means they are actively slowing the work and disrupting the chemistry of excellent team members. They have to be removed. All of this is usually painstaking and time-consuming, although necessary. It's a tough opportunity cost for any organization.

Here are a few things (red flags) to watch out for:

1. Look out for motivation beyond getting a job. People who
    articulate some additional connection to the purpose or
    mission of the company are often a better fit. But watch out
    for insincerity.

2. Look for people who are thoughtful about the PM craft and
    can articulate tradeoffs in the product process. Watch out for
    people who are opinionated about how you should do things
    without understanding the constraints your company faces.

3. Watch for new hires who start with curiosity vs. being opin-
    ionated about ways of working. Just doing what you did at
    your old job and insisting on its rightness can be a big red flag.

It might appear that larger companies that already have good product hires
would make hiring easier, but they have different challenges. Existing team
members that can gut check potential hires is a big help, but recognize that people
tend to hire people like themselves. As a leader, you should resist the urge to look
for clones of your existing team and focus on the hiring attributes you need,
especially if you can identify people with relevant experience that are different
from your existing team.

ELEVEN

Setting Direction: Systems of Strategy

Great founders have the skill of aiming their limited resources at the exact right opportunity at a given moment. Timing and aim is everything for a startup. If there's a mistake in either, the company is probably toast.

Great organizations don't rely on a single person's ability to aim. They build systems of wayfinding—of setting direction and strategy. The quality of aim and how hard everyone in the Shipyard can focus on the aim can make all the difference between success, stagnation, or outright failure.

So what are you aiming at? First are customer problems—starting with the sharpest. Then solve market problems along with customer problems (opportunities and competitive threats). Defining the aim—the product strategy—is the topic of this chapter.

WHAT IS STRATEGY?

Product strategy is the plan and approach to develop, market, and deliver software products that maximize customer delight and business outcomes. Note how congruent this is with regard to how we think of being product-led as caring deeply about the entire customer experience (which needs careful focus and curation).

It includes defining the product vision and goals, researching and understanding target markets, analyzing the competitive landscape, defining product features and user experience, and determining the most effective go-to-market plan. A strong product strategy aligns with the company's overall business objectives and helps ensure the success of the product in the market.

Product strategy is inevitably about choice. Every company has limited resources relative to its ambitions, especially early on. So it needs to prioritize. A good strategy outlines the best campaign possible given its resources (human, culture, and capital), its customers, and the competitive and market environment.

Strategy is equally what to do and what *not* to do to win. Every choice in strategy should have well-articulated opportunity costs in order to really evaluate if the affirmative choices are the best that are possible given the circumstances. So what makes a good strategy?

Good strategy usually flows from four main ingredients. The exact mix depends on the company stage, market segment, and other factors. Let's go through each one.

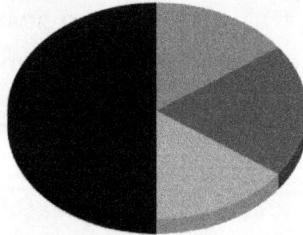

Strategy Components

● Market Environment

● Customers

● Culture & Talent

● Company Ambition

Market Environment

The first ingredient is a real understanding of what the market environment is and how it's changing in the near and long term. The market environment itself can be further broken down into trends, value chain analysis, and competitive environment.

Trends

Trends are a synthesis of where the market is going and what its important characteristics are. Trends contain insight into what is happening with your customers and how their behavior and tastes are changing. For example, the aging of the population in the US and the world and the rise of Millennials/Gen Z as the significant proportion of the population may offer unique opportunities. For a B2B software company, applying those trends to the business of work is particularly crucial.

In another example, Google's business is roughly correlated to the amount of internet access and accessible devices in the world and the growth of a literate world population (i.e., seeking information is a very basic human need). This can offer an explanation of Google's many projects to bring internet access to more people and to grow internationally.

In general, we have noticed that very successful companies' strategies are based on tying their businesses to ever-increasing trendlines. If the underlying customer trend is growing quickly and you solve it well enough to have product-market fit (and continue to expand that), there is a good chance your software business will grow to be very large.

An accurate understanding of the market environment is critical to good strategy formulation. Two other areas that should be well understood within that are the value chain and the competitive environment.

Value Chain Analysis

Value chain analysis is a management tool used to understand the flow of activities and inputs involved in creating and delivering a product or service to the customer. It was introduced by Michael Porter in his book *Competitive Advantage: Creating and Sustaining Superior Performance.*

Value chain analysis helps a company understand how it fits into the wider industry with respect to delivering its service to the customers. For example, in the current era, many software companies build their software by utilizing public clouds like AWS. Thus AWS is part of their value chain. The value they deliver is inextricably linked to the value delivered by AWS and what they pay for it.

Companies should seek to understand their value chain to make sure they identify:

a.   Economic opportunities to make more money—in certain cases you can capture more value by building pieces of the chain yourself (e.g., Amazon built out a logistics business different from USPS and FEDEX to earn more profit)

b.   Threats to their competitiveness—in certain cases you are funding your competition (e.g., Walmart does not host any of its web services on Amazon)

c.   Opportunities for cost savings—in certain cases you can deliver on a service in the value chain for cheaper (e.g., Google hosts its own cloud services)

Competitive Analysis

It's important for a technology company to understand who its competitors are, how they are exploiting the market they are in, and how well that is going relative to its own progress and strategy. Great product leaders use public and private information to synthesize their relative positioning to their competition and then articulate the various ways their products can be sufficiently differentiated in their target market.

Usually, there are different kinds of competitors and they should be categorized accordingly. In addition, if the competitors are in the same weight class (and even if they are not), product leaders should develop a credible competitive counter for each significant competitor. Failing that, they should simply acknowledge an open flank.

Calendly identified five kinds of competitors and noted the top two companies for each kind. They also developed ongoing and emergency countermeasures for each one as the insurgent market leader. At Atlassian, its Stride communication platform (the next gen replacement for HipChat) was the plucky upstart, and Oji had to develop a more detailed strategy to be differentiated in the market, knowing the strengths of all its competitors.

Target Customer

The second ingredient of good strategy is a solid understanding of your target customer—their pains and needs and how that translates into opportunity for your company.

A precursor to this is a clear articulation of the target customers. The central idea is that not all customers are desirable for your company. For example, Hermès is generally not interested in the people who shop at Walmart (jokes on them, Walmart actually sells pre-owned Hermès handbags circa 2023).

Companies usually build products for specific identifiable customers. Product leaders must find a way to describe these customers intimately and deeply for their teams. This serves as a focusing device for a range of activities: customer research, competitive research, market sizing, and, of course, product building.

Most customer "personas" suffer from a lack of depth. They discuss use case and market segment but don't understand the customer on the personal level. What is needed is a taxonomy that has such depth that it makes it easy for teams building products and taking them to market to really understand. It helps them to prioritize problem-solving for the target customers and ignore requirements that do not apply to them, to focus their time and resources. This avoids a boil-the-ocean approach, which can be quite wasteful for smaller and mid-sized companies.

Articulating target customers has to have additional depth in the following ways (for B2B):

Role

This is what they do in the company. This includes common titles and their role in the buying/acquisition process, if available. Additional color on the last few purchases they made and how they went is useful if available.

Persona(s)

This describes the typical behavioral dimensions of the target customers—things like their preferences and relative expertise with technology—and the preferences and attitudes they bring to exploring your category of software, choosing it, and adopting it.

Objectives

Objectives are the professional and personal goals and motivations of target customers. It's also very helpful to understand the customer's entire set of priorities.

Sharp Problems

These are the more painful aspects of their day and work—what gets them really worked up, to the point of wanting to bash their computer in with a baseball bat (*Office Space*-style).

Tasks: Jobs to Be Done

This is the actual job they are trying to accomplish. It also includes how they like to relate to their software and their relative sophistication at operating it. If discrete tasks within this job can be articulated, that can be helpful.

Workflows

This is the series and sequence of tasks and tools they use to accomplish their professional and personal goals at work (their key Jobs to Be Done). In other words, a workflow is the *how* of a Job to Be Done. You want a focus on the specific workflow your software solves for, but a wide lens to see how it connects to other related workflows. It's useful to understand the critical workflows that surround the workflow your product enables.

After solving for your target customer's target workflow, the next revenue opportunity is solving new, adjacent problems they have. And because adjacent problems often center the same customers that you are selling to, there are usually certain economies of scale to be reaped in your go-to-market. For example, Intuit was able to profitably expand from accounting to tax prep. Calendly was able to expand from meeting links to meeting polls. Atlassian expanded from developer workflows to product management workflows, etc.

Company Size and Industry

B2B firms should identify the kind of company your target customer works in. Is it a small-sized company, mid-market, or large company? Company size helps you approximate their needs and budget. What specific sector of the economy are they in? What is their typical revenue? A clear firmographic sense assists tremendously in go-to-market efforts across the board.

Target customers should be initially defined as narrowly as possible without making the implied market unviable (i.e., greater than $20–$50m TAM). A product that is for a certain kind of healthcare professional is better in this sense than a product for every healthcare worker. The latter description is likely not apt for the first few generations of your product and is a disservice to your marketing employees.

Below is a simple set of steps for segmenting and describing your ideal customers, especially in the early stages. Although this also works for later-stage B2B software companies. Consumer software companies may need to do additional work because of the scale and variety of their customers:

1. Look in your company product-usage database and find the
   customers who use the product most frequently.

   a. Develop high-use AND high-satisfaction metrics to aid
      in your search.

   b. Factor in conversion time. Some customers who found
      you early and are succeeding may not be the same people
      succeeding more recently.

2. Cluster them by any JTBD nomenclature you have gathered
   via onboarding.

3. Survey a broad sample of these customers and ask them about
   their companies, themselves, and their use cases/workflows.

   a. Re-cluster them by their own described logic and see if this
      matches the one you used for your onboarding.

   b. If it doesn't, make a note to change it so it captures real-
      world JTBD.

4. Interview a subset sample to really map out their company
   types, motivations, and workflows deeply.

5. Make sure these customer/company descriptors feel strong
   enough to build a business on (i.e., we feel confident we can
   understand them and market to them in the millions/billions
   [SMB] or in the thousands [enterprise]).

A complete and ongoing synthesis of the target customers is very important to
picking the right strategy. A well-defined target customer set helps tremendously
with aligning time and resources to the right initiatives while avoiding activities
that waste time and effort on unprofitable customers.

It's important for any software company to have an ongoing synthesis of its best
customers and what problems, if solved, will ensure their delight and loyalty. It

is important in everyday execution but even more important as you direction set and develop and evolve your company or product strategy.

Culture & Talent

One definition of strategy is leveraging a company's competitive advantages and unique strengths. That's where culture and talent comes in.

Culture and talent don't just define a company's potential, but their unique superpower. We've seen companies whose advantage lies in their particular talent, such as a strong AI dev team, a uniquely talented sales team, or a culture that works and ships quickly. In each of these cases, these things should be considered when applied to strategy. Differentiated talent leads to valuable assets, such as hard-to-build technology, brand equity, or a viral product. These assets are often the direct result of talent in the company.

Sometimes it's a more subtle blend of things often referred to as culture. For example, Nintendo understands how to make both great consoles and great games that can thrive regardless of the competitive pressure they face in the market (from 3DFX, Sony, Microsoft, Sega, etc.). This allows them to innovate and thrive even under the harsh glare of two well-funded competitors like Microsoft and Sony. Their gaming culture has sustained them over the years and capitalizing on it is absolutely part of their strategy.

Company Ambition

The fourth ingredient of strategy is the ambitions of the company. What kind of company does it want to become? What niche does it want to occupy? And what kind of business model does it need to devise?

Companies can decide to serve consumers, business customers, or developers. They can decide to build picks and shovels instead of mine for gold—in other words, build a tools company or a platform company. A good example of this strategy is Plaid, a fintech company that helps banks and payment tools integrate with one another. Plaid enables online payments and gets a small piece of billions of transactions. PayPal, on the other hand, only makes revenues from transactions that happen within PayPal.

A key part of a company's ambition is the business model it develops and how its revenues accrue. For example, Procore Technologies's ambition is to be a partner in the building industry's growth. As a result, it charges based on a percentage of the total capital costs of a building project. This model is also preferred by its customers—general contractors and development companies—given the unique way building projects are financed in many countries. Compare this to their top competitor, Autodesk, which charges a subscription. Autodesk's ambition is scale, which they have done quite successfully. Autodesk is many times larger than Procore, while Procore generally has more favorable reviews from customers.

Now let's return to the question at hand: What makes good strategy? We have identified its components, but that does not yet stipulate what *good* looks like. Good strategy involves evaluating your inputs, determining your strengths and challenges, and then selecting credible levers that will drive growth and competitive advantage. We address growth levers and competitive moats next, but first, let's discuss a few ways that strategy for companies at different stages differ.

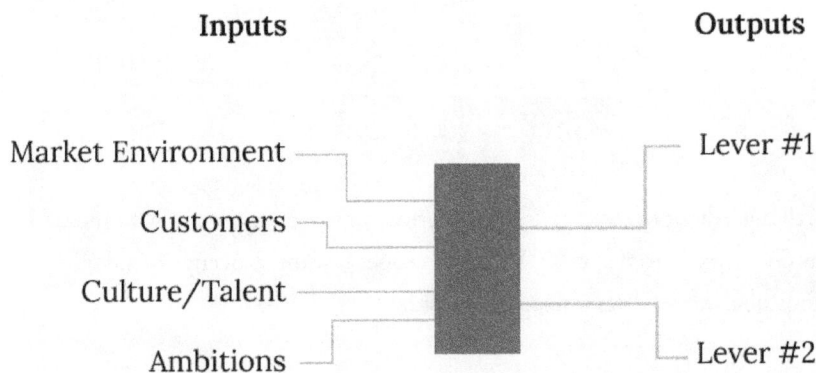

Inputs		Outputs
Market Environment		Lever #1
Customers		
Culture/Talent		
Ambitions		Lever #2

STRATEGY AT EACH PHASE OF GROWTH

Our good friend Gibson Biddle, former Chief Product Officer at Netflix, defines product strategy as the answer to the following question: How will your product delight customers in hard-to-copy, margin-enhancing ways?

Product strategy evolves as a company grows. What works for an enterprise will not be the best path for a startup and vice-versa. Here is how we think about product strategy at each phase of growth, starting with new ventures.

Startup Phase

Startups have no real strategy to speak of. They are a bit like small, feral creatures who are searching for a repeatable business model. While they consider the trends, competition, and customer targets they are focused on, their main focus is identifying a sharp problem for a specific customer base.

Another way to put this is that a startup is the process of discovering a coherent and profitable strategy—the work of a startup is primarily customer discovery and translating those insights into a working and valuable product. The vast majority of a startup's strategy should be focused on discovering, understanding, and marketing to their target customer.

Startup teams should be wary of devoting too much bandwidth toward big S strategy, unless it supports finding product-market fit. Otherwise, it can be a waste of valuable time. In a startup, the elements of strategy function more as insights that indicate an opening in a specific market that can be exploited at a cost and speed that is hard to match by any incumbents.

Mid-Phase

Companies that are scaling start to have a more formal strategy because with some success comes operational concerns and competition. However, the most valuable element is still the ability to satisfy customer needs innovatively and rapidly. As a result, customer concerns and needs dominate other elements of strategy by a good margin even at this stage.

The best companies start to become aware of the competition and use that awareness to make product choices that make them differentiated while pumping out hard-to-copy investments that still meet customer needs. This is not exactly a moat, but it's an effort to run up the score—make yourself both unique in the marketplace and also difficult to compete with or catch up to.

Scaled

At scale, companies have to consider many more elements of strategy—really, the full gamut. Often there are multiple products, multiple distinct organizations, and a plethora of concerns, including regulatory, competitive, environmental, and more. Size by itself is a challenge because eking out real growth implies the marshaling of many resources to solve novel problems or attack novel markets. Strategy is even more complicated for public companies with the constant expectation for growth. (Note that this is also similar for venture-funded companies.)

The pressure of size means that strategy is often dominated by many concerns that are not the customer. Therein lies the main challenge for most sizable companies: continuing to focus on their customers while they grapple with all the other issues they need to pay attention to.

Great product leaders in larger companies continue to prioritize customers alongside their other concerns. Not doing so risks decline and being disrupted by more customer-focused companies. Insurgent companies tend to defeat scaled companies because they have a better ear for customer needs and have crafted far better solutions in response to that.

Enterprise Transformation

Enterprise transformation is a unique strategy challenge. Unlike startups or scaled companies, which are often building from a blank slate, enterprise transformation involves navigating legacy systems, entrenched processes, and established cultures. As Ezinne's experience at Time, Inc. illustrates, successful transformation requires a thoughtful, disciplined approach to strategy.

At the heart of Ezinne's approach is a 5–7-year company strategic plan. This long-term vision simply serves as a guidepost for the transformation journey, ensuring that all initiatives and decisions are aligned toward a common goal. For Time, Inc., that goal was to transition iconic print brands like *Time*, *People*, and *Sports Illustrated* into profitable digital businesses.

To make the strategy actionable, break the long-term vision down into three-year chapters, each with its own specific objectives and focus areas. These chapters could include re-platforming or going global. In addition, enterprise transformation often requires dismantling or replacing existing infrastructure. This can be a delicate process, as it involves challenging the status quo and disrupting established ways of working.

At Time, Inc., the shift from print to digital required a fundamental reimagining of the company's products, revenue models, and organizational structure. To manage this transition, Ezinne had to work closely with her counterparts across the organization to build alignment and buy-in. This involved clearly communicating the rationale behind the changes, as well as the benefits they would bring to the company and its customers.

Cultural change is another key aspect of enterprise transformation. Established companies often have deeply ingrained behaviors and mindsets that can be barriers to change. At Time, Inc., the transition to digital required a shift from a content-centric culture to one that prioritized user experience, data, and agile experimentation. This is where people and management systems are crucial. Smart hiring (and firing) and incentive structures are key to changing behaviors and shifting the company culture.

Enterprise transformation is a marathon, not a sprint. It requires a clear vision, a flexible roadmap, and a willingness to continually adapt and evolve. By breaking the journey down into manageable chapters, and by fostering a culture of adaptability and customer-centricity, product leaders are able to guide their organizations through even the most complex transformations.

GROWTH LEVERS AND COMPETITIVE MOATS

There are many frameworks and books on strategy, both for the whole company and for the product. In technology companies, most of the strategy *is* the product strategy—the things that draw in customers for their use/benefit. However, a few other things matter significantly: How the company sells and how it goes to market, especially early on.

We've studied strategy at top business schools and, more importantly, in the trenches of fast-growing technology companies. Tech companies have been around long enough—about fifty years of concentrated history—that we can now study what works and what doesn't with real clarity.

Through this study of strategy and applying it in large technology companies, we've come to understand that two strategic frameworks stand above the rest in terms of impact: growth levers and competitive moats. We've dedicated a chapter to each, coming up next.

TWELVE

Growth Levers:
Driving Reach and Value

There are finite levers for the growth of a product-led company, especially in the early stages. These levers can be broken down into two main categories: those that increase the reach of the product and those that increase the value of the product.

Increasing the reach of the product is about making big leaps in bringing essentially the same product to more of your core audience or expanding the audience to new customers.

Increasing the value of the product is about adjusting or improving how it solves the target customer's workflow(s) or whether it grows to solve additional workflows that were not previously targeted.

There are ten growth levers every product leader needs to know:

Reach Levers:

- Increase activation and conversion rates
- Increase virality
- International expansion
- Channel partners and re-sellers

Value Levers:

- Improve the existing core application
- Expand the value proposition to new audiences
- Sell upmarket
- Sell downmarket
- Build a developers platform
- Pricing

Obviously, every growth lever comes with an opportunity cost. Use your go-to decision-making framework to evaluate each of these levers. We prefer a simple ICE evaluation: Impact, Cost, and Effort.

Levers of growth for B2B SaaS companies	Category	Impact	Cost	Effort
Activation/conversion/growth hacking	Reach			
Increase virality	Reach			
International expansion	Reach			
Channel partners	Reach			
Improvements on the existing UX / core app	Value			
Increase value prop. that attracts new audiences (absorb more JTBD)	Value			
Upmarket (Prosumer > SMB > Enterprise)	Value			
Downmarket (Enterprise > SMB)	Value			
Developers (i.e., build a platform)	Value			
Price	Value			

Let's look at each lever in more detail.

REACH LEVERS: EXPANDING THE CUSTOMER BASE

Improve the Activation and Conversion Rates

Many products have a tremendous amount of opportunity for increasing their activation and conversion rates. Everything from adopting a self-serve flow (where it makes sense, which is most cases) to optimizing your trial model (trial, freemium, reverse freemium, etc.) to optimizing onboarding, streamlining pricing, etc.

By optimizing the customer journey in the first fifteen to thirty days, product teams can increase their ability to activate and convert more customers. However, it's important to note that there are a couple of assumptions here: Your product has accomplished product-market fit for some segment of your target market and you have a reasonable way to get some of your customers to find your product (go-to-market). See chapters 4, 5, and 6 for specifics on streamlining customer activation and conversion.

Increase the Virality of Your Product

Many viral techniques can be added to your product over time. Virality means less work for your marketing team if you are able to harness this engine properly. Viral loops by themselves tend to not be the highest-performing acquisition technique, unless your product has broad appeal. Part of the reason is that viral loops tend to cast a wide net and attract a lot of customers for whom the use case does not apply or who are not yet convinced to buy (broad appeal counteracts that). For example, a lot of people who encounter a viral sales tool may not be in the market for a sales tool. However, given that a viral loop is essentially "free" marketing, it's better to have one than not, even a low-performing one. And as your product scales, the dividends of a viral loop increases (i.e., K-factor in the context of much larger numbers of referrals).

Our earlier discussion of synthetic vs. durable virality in chapter 8 details useful methods to increase the viral reach of your product.

International Expansion

Really great products tend to grow internationally organically. However, most have a major market—usually the origin geographic market and any other adjacent geographies. For companies that originate in the US, most of their revenues come from the US but also English-speaking countries like Canada, Australia, the UK, etc. This effect is approximately because marketing spend tends to be regional, and thus so does audience building. Sales operations and support operations also tend to be regional. Payments can add other friction in regards to pricing in foreign currencies, but so does dealing with different regional tax jurisdictions.

Expanding operations internationally, therefore, is non-trivial but often worthwhile for scaling companies. It usually starts with localizing the product (language translation, regional features, etc.) and continues with some sales and marketing presence in key markets.

Channel Partners

Sales and marketing expenses are often significant for scaling technology companies (less so if you're product-led, but still substantial). If the conditions are right and the incentives well-aligned, technology companies can opt to work with channel partners and resellers to amplify their sales and marketing reach by creating mutually beneficial arrangements.

The key to successful channel partnerships is strategic alignment. Partners must have complementary goals and capabilities, with clear value creation on both sides. In the mid-2010s, T-Mobile recognized that customers needed WiFi access outside of their homes and offices. Ezinne, who was the product leader for the T-Mobile Hotspot @Home initiative, explains: "We focused on the 'third place'—the venues people frequented most, like coffee shops and trains."

T-Mobile approached Starbucks and Amtrak with a win-win proposition. T-Mobile would provide the WiFi infrastructure while the channel partners provided the venues and customers. This allowed each company to focus on its strengths while tapping into the other's resources.

The partnerships were hugely successful. Starbucks became one of the largest WiFi networks in the country, and T-Mobile gained a powerful marketing asset. Customers began associating T-Mobile with connectivity on the go, strengthening the company's brand.

When considering partnership opportunities, look for allies that share your objectives and can help you reach new audiences. By leveraging the right partners, you can accelerate growth without losing focus on your core competencies.

VALUE LEVERS: PRODUCT INITIATIVES THAT DRIVE GROWTH

Improve on the Existing Core Application

The biggest value lever for a software company or product line is to improve the core application. This can be an improvement in the user experience (a design investment) or additional functionality (a multidisciplinary investment).

The idea of product-market fit comes into play here, especially early on. A really simple, powerful, and well-designed application can be used to build a strong business. However, it's a journey to find customer opportunity and deepen it over time through building new things. In addition, companies often find that when they achieve PMF, they start to notice adjacent opportunities that serve the same customer that they can use their existing expertise to serve efficiently.

Many software companies (e.g., Apple, Microsoft, and Adobe) have an annual cycle of releasing new features aimed solely at improving their core applications. Many cloud-native companies release features more frequently in order to win new customers.

When a company is at scale, improvements in the core application can stall as a growth strategy, depending on how deep the addressable market is and the competition level. Part of the reason is that complexity in the product tends to increase over time. New features make a smaller incremental impact and are harder to discover. In addition, marketing efficiency decreases, and driving excitement over new functionality wanes. In this case, building or acquiring new products starts to play a larger role in a company's growth strategy.

Expand Your Value Proposition to Attract New Customers

Technology companies can also opt to build something for a new audience/customer. An example is building a tool for salespeople and then adding functionality that appeals to marketers as well (or in the case of HubSpot, just the opposite).

There are some risks with building for new customer segments that must be noted. A key decision is whether a brand new product is needed to serve the new customers customers vs. an extension of an existing product. If, however, it is determined that a new product is necessary, then you will have to pay for additional research and development up front. Does your team have the bandwidth, or will you have to add more people? Finally, will this new product take away attention from your core product? There is a risk of splitting the attention of your team between products and hurting both.

Building new functionality into the same product has risks of adding complexity to the core product, though the upfront R&D costs will be lower than building a brand-new one.

An efficient way to expand into new value propositions is via acquisition. However, this trades some of the risks of building it yourself (including the time it takes to build and mature new functionality) for different risks. It takes time to integrate new products and the team(s) building them. It also tasks your existing team with not only figuring out an integration strategy, but also learning how to build and support a brand new product.

Go Upmarket into the Enterprise

Many B2B SaaS companies enter the market at the lower end by targeting small and medium-sized businesses (SMB). This is very common for many reasons. These companies often have lenient procurement standards and are amenable to bottom-up demand for tools, and thus a product-led growth strategy. They are also easier to sell to via marketing and SEO vs. larger companies.

However, as a company scales with these techniques, another path to growth is to sell to larger companies. This usually comes with a different set of challenges:

- A sales team that can sell to the enterprise

- New compliance requirements often demanded by enterprises

- A strong financial position that means that the software service will be around for the long haul

- Changes to the release cadence to make sure that big companies can digest the pace of a typical cloud-native company

In many ways, moving upmarket means developing an entirely new product, or at least a new version of your existing product. As a product leader, you must also consider the staffing required to support enterprise-level customers.

Of course, the opportunity is to drive higher-priced and higher-volume deals, which means that a software company can grow its revenue much faster on the back of a smaller customer base. As a result, perfecting an upmarket motion after building an SMB market is a significant growth opportunity.

Go Downmarket

One downside of building a software business for enterprises is the competition. There is no shortage of vendors trying to sell into this very attractive market, so the competition is fierce—one doesn't only need to have a good product, one must stand out. In addition, sales cycles are longer and there is a higher dependence on highly compensated salespeople. And any inevitable churn will represent big chunks out of a company's revenue flows.

In certain cases, enterprise software companies may choose to move downmarket to sell to smaller, more plentiful businesses. Bazaarvoice, a leading provider of product reviews and user-generated content solutions, had historically focused on serving large retailers like Walmart and Target. Enterprise retailers were only interested in the top brands that drove the majority of their sales, but there was a massive long tail of smaller brands that were hungry for the same review capabilities. Ezinne led the company's expansion into supporting the tens of thousands of brands that populated the retail shelves. Smaller brands, while lacking the purchasing power of major retailers, made up for it in volume and growth potential.

As with going upmarket, moving downmarket requires a rework of your entire product strategy. Bazaarvoice needed to adapt its offering. They developed a simplified, self-serve platform with affordable SaaS pricing, allowing even the smallest brands to collect and syndicate reviews. They had also built out a review syndication network, enabling brands to distribute their reviews across multiple retail sites as they acquired distribution deals.

The shift required operational changes as well. Bazaarvoice invested in automated onboarding and support to efficiently serve a high volume of smaller customers. They also realigned their sales and marketing to target individual brand managers rather than retail executives.

The move downmarket paid off handsomely. Bazaarvoice rapidly grew its customer base, adding thousands of new brands. The revenue from these smaller accounts, while individually modest, aggregated into a significant new stream. Bazaarvoice had tapped into a well of latent demand, and in doing so, diversified its business and accelerated its growth.

When well developed, SMB revenues can be quite resilient because they're spread across many more businesses. However, the challenges can be daunting—even more so than selling into the enterprise. Selling to SMBs is a special skill set for any company. In addition, doing so only works at scale. The essential bet is that your product will be adopted by an order of magnitude more SMB customers than enterprise consumers and a much lower price point. It's a tradeoff between price and volume.

Because most enterprise-focused software companies aren't structured to easily to move downmarket, this is often the vector of competition from insurgent companies. For example: Figma (small teams and freelancers) vs. Adobe (enterprise), and Trello (small teams and individuals) vs. Atlassian (medium-to-large teams).

Smart enterprise-focused software companies often consider acquisitions to shore up the low end of their market to get started and to protect their flanks, in addition to driving incremental growth. Others have little choice but to retool and make this move to survive because of how challenging the enterprise market can be.

Build a Platform

Every software product can become a platform. At its most basic, this means that some or all of its functionality can be accessed by a developer via simple application programming interfaces (APIs), allowing others to use your technology to build their own workflows.

Becoming a platform has several benefits. The first is that for the subset of customers who use your API, your product becomes immensely sticky. Once your product is used in their workflow via code, it's hard to switch to a different vendor. This is an efficient way to increase customer lifetime value (LTV) and maximize switching costs.

Second, if a software company can attract thousands or millions of developers to build on its platform, it gains multiple kinds of scale, including network effects. If developers spread the word about your platform and take it upon themselves to create content that helps other developers adopt your tool, your growth will skyrocket. In the best-case scenario, your platform becomes the go-to choice for developers and your tech becomes critical to the business models of other companies. For example, many software businesses are entirely built on the Atlassian platform or the Shopify platform. The key to reaching this pinnacle of platform design is opening up an application marketplace and featuring the companies using your tool.

Building a platform is not without its risks. The developer customer is usually very different from the typical target customer a company serves; it takes time to understand them and then monetize them. Usually the business model is entirely different from selling to a typical worklow-focused customer. Still, very few large companies in software history have NOT been platform companies. Facebook, Twitter, Microsoft, Google, Amazon, Shopify, Atlassian, and countless others became world-changing companies this way. It's an endless roster. A platform is all but necessary to reach the highest echelons of business growth.

Adjust Pricing

An underrated value lever for growth is pricing. On one hand, a company can simply raise prices on its products when it thinks it has pricing power and the volumes will not decrease linearly. Or, if a company has a robust sales channel, it can reduce the pricing to sell more volume. There are multiple levers within the pricing topic to pull: monthly vs. annual pricing mix, pricing and packaging structure, the pricing user experience, and more.

Here are some words of encouragement for product leaders: Pricing is not a delicate lever. All software companies should regard pricing as an area of active experimentation and ongoing optimization to increase the growth of their business. At a certain scale, every software company should employ people whose full-time job is to optimize all aspects of pricing continuously.

As you design your strategy, remember that every strategic choice ripples through your entire organization. The growth levers you choose to pull will determine the talent you need—whether that's pricing analysts to optimize monetization, platform engineers to build your ecosystem, or global expansion specialists to drive international growth. Your strategy shapes your hiring, which in turn determines your Shipyard's ability to execute. The most successful product leaders understand this interplay and design their strategy accordingly, ensuring each decision strengthens rather than strains their organization.

If you study carefully, these ten growth levers cover the majority of growth strategies for the last forty years of technology company progress. We've covered them at a high level to raise your awareness of them, but each lever deserves deeper study—there are entire books written about individual levers like pricing strategy or international expansion. We encourage you to take time to thoroughly understand the levers most relevant to their strategy before pulling them.

Growth levers are important tools in every product leader's tool kit, but they are many times more effective when selectively applied to the right opportunities. The point of growth is to make it endure, and the only way to do that is to make sure the growth you invest in can also confer sustainable competitive advantage. So let's talk about how to systematically develop competitive advantage next.

THIRTEEN

Competitive Advantage:
Building Moats for Product-Led Companies

Competitive advantage is about building a durable, hard-to-copy, high-margin software business that also grows quickly.

Companies that develop competitive advantage generally have an edge in earning profits in any given market. They also have a chance to build a dominant market position. A company that wants sustained profits and long-term growth must seek to develop competitive advantage in order to achieve this. Simply having more revenues and customers at any given point in time is not enough.

Developing competitive advantage is not accidental. Fifty years of technology company strategy research has shown that you can develop this consciously through critical investments. One of the most impactful strategy classes we ever took was a course in technology, media, and telecoms (TMT) strategy at UC Berkeley's Haas School of Business. There we encountered a well-researched evaluation criteria for whether your technology company is on track to develop competitive advantage. This checklist helps you set paths that are intentional about your sustainable growth. There are at least ten dimensions through which companies can create competitive advantage. This list was developed by Reza Moazzami, who has a PhD in electrical engineering and computer science and also holds an MBA from MIT and taught strategy for technology, media, and telecoms markets. Each one of these dimensions by itself can be defeated by competitors. The task of any software business is to develop multiple advantages and build a strong competitive moat.

We will cover the list in this chapter to help startups, product leaders, and executives chart a path that reinforces their product strategy. In the parts that follow, evaluate your competitive advantage by scoring your company on a scale of 1–10 across each dimension. 10/10 is having a virtual monopoly along that dimension; 1/10 is having zero advantage. Anything over 8/10 is a very strong advantage, and under 5/10 means there is more work to be done.

Item	Score (1-10)	Notes
Intellectual property		
Brand		
Distribution/market presence		
Economies of scale		
Economies of scope		
Vertical/horizontal integration		
Network effects		
Access to capital		
Regulation		
Competition		

Pro Edition: Competitive Advantage Scorecard

Dig Deeper: You can review this scorecard template and see a real-life example in the premium edition: productmind.co/brpro.

Generally, a company should tackle the lowest-scored advantages first. Every low score indicates a potentially critical weakness.

Let's talk about each dimension in a bit more detail.

Intellectual Property

Intellectual property (IP) in the context of technology and software companies refers to inventions, software, processes, algorithms, and creative technology that is unique to your business and used to create its product and services. Because competitive businesses are prone to copying, IP has to be protected by secrecy, patents, trademarks, and copyrights to be asserted as an intangible asset that can add to the value of your business. IP, when properly protected, can be used to prevent trivial copying of your business by competitors through mechanisms of the law. IP is also an asset that can be assigned a financial value by an accountant. As such, it's tradeable for financial value, but also as a defensive asset. Competitors will often sign cross-licensing agreements when their IP prevents each other from entering critical markets where they can assert crucial IP rights. In fact, this defensive use is the main application of many large software companies when they think of generating critical intellectual property.

Intellectual property is a fraught topic in the technology industry. Many feel it should not apply to software and algorithms because it creates a chilling effect on innovation and an unfair advantage for incumbents. People also argue that many inventions that are considered IP are actually trivial and should not be protected by governments. The open source movement, which has become so valuable in the construction of modern technology businesses, is in many ways a repudiation of the global intellectual property regime as it stands today in relation to software.

Our view is to sidestep this issue entirely and assert that every software business should use every legal lever to secure not only its success but its advantage, within ethical boundaries. IP can add significant value to the terminal and exit value of a software company. It's also a magnificent defensive weapon when there is IP from another company that prevents you from innovating. Imagine for a second that another software company—even a competitor—has a crucial technology, for example a codec that could save the target market a billion dollars every year. Your company, on the other hand, has IP that could save the same market $500m yearly. In this case, instead of litigating whose technology was invented first and is better, your company could strike a licensing deal that has more generous terms than it would have gotten if it had no IP to assert.

In the best case, your company might have so much talent that you have unique solutions to hard problems that no one has ever imagined. Benefiting from that long enough to capitalize on it is the very essence of developing competitive advantage via IP.

Qualcomm has built an incredibly strong business on its cellular technology patents, particularly in 4G and 5G technologies. Their IP portfolio is so extensive that virtually every smartphone manufacturer must license their technology, generating billions in high-margin licensing revenue. This IP advantage has helped Qualcomm maintain its dominant position in mobile communications for decades.

Brand

A good brand can create incalculable value to any company, but particularly a technology company. We have had the privilege of working for some iconic companies with great brands—Microsoft, T-Mobile, Time, Inc., Atlassian, Twitter, and Calendly. These companies had easy-to-love names, brand marks, and more. A great brand starts with making successful products, but beyond that are the deliberate efforts of a competent marketing team.

In technology company circles, we often talk about the point where a brand becomes a verb, for example, Google or Uber, via market category dominance. This is a debatable point and there are plenty of counter examples; just think Apple and Microsoft. However, strong brands can make up for a lot of missteps in a software company's journey, which creates a competitive advantage. For example, Uber's brand helped it endure a round of municipal and state rejections in the 2010s. A few municipalities banned the ridesharing app (along with its competitor, Lyft) when they refused to comply with "onerous requirements." One of those cities was Austin, Texas, where we live. For almost two years, local ridesharing companies filled in the market gap, but when Uber (and Lyft) came back to the city, they rapidly dominated the market again. Their brands were strong, international, and associated with quality. This is the definition of competitive advantage.

In some cases, customers will wait for a reputable brand to deliver a feature offered by an upstart with no brand, especially if the trusted brand promises to offer it. This is counterintuitive because the market *should* reward the company first to market. But in this case, a good brand from an alternative company (in conjunction with a well-timed announcement), can disrupt the ideal of a perfect market.

When does brand become important? Once a software company has established product-market fit and shows growth, it should turn its attention to building a recognizable brand. Domain name, brand name, brand marks, brand marketing, etc. should all be invested in to make sure that it can start to build competitive advantage in this fashion.

It's worth noting that a good brand is not an insurmountable weapon. A new entrant that delivers on a customer promise 5x better than the old promise a company has made to its customers can be disrupted. At that threshold of workflow augmentation and acceleration, customers are prone to switch horses. However, a great brand can keep the switching costs threshold high. It very well could be that a 3 times improvement in productivity that topples an ordinary company with zero brand investment is not enough to do so for a well-branded company.

Distribution and Market Presence

Establishing an advantage in routes to market creates competitive advantage. A simple example: If you can afford to pay for a Super Bowl ad (and it's effective for your market) and your competitors cannot, you have an advantage in distribution and market presence. More relevant to software companies: If you can afford to pay for performance, digital marketing, and direct sales for your software products and your competitors cannot, you establish a sustained advantage. There could be a range of reasons your company is able to do this—more capital to spend, higher LTV of your customers, lower customer acquisition costs for your customers, etc.

In practice, eventually, all routes to market can be replicated by competitors with large enough scale and partnerships. So software companies have, over time, developed the practice of performance marketing—the ability to have visibility and engage all routes to market as a marketing mix and optimize all the levers to maximize the impact of marketing and sales spending. Performance marketing is a growing field and many approaches are still proprietary. This means those that do this better can eke out an advantage.

One major lever a technology company can pull at scale that many do not have is effective reseller relationships that can amplify your sales investment. Resellers are third-party businesses that are officially licensed to sell your product and receive a royalty or share of revenue in return. This channel is highly effective at scale, but is difficult to perfect and requires significant investment. A top-notch sales team is also a great investment to make if it's appropriate for your product and market segment. for example, Microsoft's enterprise sales force conveys significant advantage in selling to the enterprise.

Companies like Salesforce have gained market power through some incredible investments in distribution and market presence.

Companies like Salesforce have gained incredible market power through investments in reseller partnerships and sales. They have mastered indirect distribution with resellers, affiliates, and managed service providers in addition to their own sales force. Salesforce has created an army of other companies whose success depends on the expansion and continued growth of Salesforce, and this virtuous cycle feeds into the company's other competitive moats, such as economies of scale.

Economies of Scale

The idea of economies of scale is simple: The more of something you make, the cheaper many input costs can become on a unit basis. The more you make, the more your negotiating power grows over your suppliers, which can further drive down costs. The benefits of preferential discounts and volume pricing can drive down marginal costs, which can then increase marginal profit. Economies of scale first showed their might during the Industrial Revolution. In the software world, they are more powerful than ever.

Unlike physical goods such as cable, real estate, or railroads, the fixed costs of building and running software is a fraction of the total cost, at scale. Adding one new customer costs virtually nothing, meaning software companies have come to enjoy profit margins of 80–90% (this is not quite as true if your company also makes hardware; in that case, it will be more in line with most traditional companies). Even the marginal costs of running software become cheaper at scale. Famously, Google and YouTube do not pay for egress bandwidth like most software companies. Their traffic is so large that they use peering arrangements with other large internet backbone providers and ISPs to reduce their operating costs.

In other words, the scaled growth of software companies are only effectively bound by demand, not costs. This is why product-led growth is such a powerful idea; if you can reach economies of scale without spending massive amounts on sales or marketing, you can become a unicorn many times over.

Developing economies of scale in the software business is a very important milestone. Which raises the question: What does scale mean? Here are a few scale dimensions to consider:

- Repeatable revenue over a year: The ability and business model to generate sustainable revenues of over $500m year over year. For reference, the top fifty companies on the NASDAQ have an average annual revenue of $50b circa 2021.

- Access to public markets: The reliable ability to raise debt and equity financing at current market rates.

- Scale in geography: The ability to address demand in world-wide or near-worldwide geographies.

- Scale in people and operations: The ability to have the talent and operations to tackle multiple opportunities in your target market.

Economies of scale are even more powerful than people often realize. It is not just the lower input costs and preferential treatment from suppliers. Very large-scale software companies (especially publicly traded ones) enjoy many other advantages.

- They can build new products in addition to the one they started with.

- They can generate more intellectual property (legal costs become a small fraction of profits).

- They can spread fixed operational costs over large volumes of sales.

- They can make major investments in R&D and create more innovative products (for example, Microsoft recently invested $10B in OpenAI and secured a highly privileged position in a new software frontier).

- They can buy out the competition and stay in business longer as a result.

- They can influence public policy related to their markets.

- They can make major mistakes in product development and survive the fallout as long as they correct it quickly (for example, Microsoft survived the debacles of Windows Vista and Windows 8 because of both scale and network effects of their Windows platform).

The above are just a few of the advantages, but note that there are disadvantages to scale too. Mainly the companies' inability to move quickly in the face of opportunity or competition. But the goal of every ambitious technology company should be to reach this level of scale and enjoy its privileges while being ethical, avoiding monopolistic tendencies and respecting market rules.

Economies of Scope

Economies of scope refers to the ability of a software company to use the same or similar resources and operational architecture to generate additional revenues.

A company can use economies of scope to build or acquire additional products that serve similar customers by addressing different needs they have by applying similar or same talent (which can thus be similarly managed), tools, and operational expertise. Companies can also apply similar go-to-market resources to multiple products in complementary ways that do not require complete duplication of resources, thereby saving on unit costs for outsized additional benefits and revenues.

The rule of thumb for developing economies of scope is to look for novel win-win collaborations with people and teams inside and outside your company. This could be other departments, teams, companies, your customers, channel partners, or developers. Here are some examples:

- Amazon's investment in a web services architecture allowed them to build a more resilient online marketplace but also to enter the cloud services business at scale.

- Microsoft's investment in productivity tools allows them to build, bundle, and acquire multiple products in the productivity space (even beyond core Microsoft Office) and derive significant additional profits. Owning the largest OS on the planet has made them the ultimate platform for other builders to co-create alongside Microsoft's core products.

- Uber's expansion to Uber Eats and Uber packages means they use the same cars, same drivers, same software investment to serve new customers and markets, etc., all serving slightly different customer personas using a similar core workflow.

- Facebook's core social platform was used to launch many related products: communities, messaging, gaming, marketplace, and more.

The most straightforward path to economies of scope is building an internal or external developer platform that abstracts the core pieces of your software. This allows different "projections" of your software to be built quickly and efficiently by the people closest to your customers. Companies with strong economies of scope also think carefully about branding. By extending their brand to multiple features or products that serve similar or adjacent customers, they are able to expand revenues from an installed base to another more efficiently.

Vertical Integration

Vertical integration happens when a company develops expertise on multiple levels to reduce its reliance on third parties. Samsung, for example, not only designs its own products, but also makes the software, CPUs, memory, LCD screens, and chips that go into the products (Samsung also has a side business of selling its extra fab capacity to other companies to be used in their finished products).

Vertical integration is common in hardware-centric companies. It's valuable when the product is very exacting, and proprietary techniques can yield additional benefits. For example, Intel owns chip fabrication plants and also designs its chips. (A peculiar anti-example is Nvidia, which only designs its powerful chips but outsources the manufacturing to trusted partners.)

Vertical integration is less common for software-centric companies. The reason is that software by definition gains power when it can exchange information with other software built internally or externally. This is why Apple's closed Macintosh ecosystem languished in the open world of Microsoft in the 1980s. It wasn't until Steve Jobs embraced cross-compatibility (and later, the developer ecosystem) that Apple's beautiful software became mainstream.

However, software companies at large scale will dabble in vertical integration where it makes sense. For example, Facebook, Amazon, Google, and Microsoft commission their own CPUs for AI compute in certain cases. A subset of these same companies design their own data centers and data center management software. Facebook, Google and Apple develop custom AI that enhances their various products and hardware (Pixel phones, iPhones).

Efficiencies derived from vertical integration can be quite formidable if it has high value and is proprietary. However, any proprietary technology has to evolve as fast as or faster than market forces can move, or else vertically integrated companies will fall behind.

WP Engine has built their strategy around the vertical integration of website hosting, developing their own solutions for every layer of the stack from infrastructure to developer tools to customer support. Rather than cobbling together third-party solutions like many hosts, WP Engine controls the entire WordPress experience, from their custom caching technology to their security tools to their development workflows. This integration enables them to deliver superior performance and reliability, resulting in industry-leading customer satisfaction rates despite charging premium prices in the competitive hosting market.

Horizontal Integration

Horizontal integration describes when a company stays in and expands into its preferred layer of the value chain through new products, acquisitions, or partnerships. A good example is Atlassian, which has ambitions to dominate work management and collaboration. It owns Jira, its developer-centric product, but also other tools for product managers, designers, IT helpdesk, and executives. Atlassian also acquired Trello, which is designed for SMB-level work management.

Most software companies will seek growth through horizontal integration. It is generally easier to become good at one thing, such as work management software (Atlassian) or building laptops (Asus), and expand to reach tangential markets that also need those things. Horizontal integration leads to economies of scope, which provide the benefit of an expanded revenue base without increased operational costs. Again, branding becomes quite important here and can lead your company naturally into new markets.

Excessive horizontal integration can lead to higher exposure to regulation, especially in small markets, because of consolidation and the acquisition of excessive market share. For example, if a company buys up all the companies in disk scanning or legal data search, it might be subject to more regulatory scrutiny because it has come to dominate that market.

Network Effects

We discussed network effects at length in chapter 8. Rightly so, because it's the holy grail of product-led market growth and is almost as immutable as gravity.

In our experience, network effects are by far the most significant competitive advantage to acquire. They can create a barrier to entry for competitors, as customers are more likely to stick with a software product or service that has a large and active user base. Additionally, network effects can lead to a virtuous cycle of growth, where more users attract more users, increasing the value of the network over time.

The most important thing to know is that network effects can be engineered. As we discussed earlier, when the product team at Calendly made it easier to schedule meetings with other Calendly customers, you could instantly see where your availability lined up. This was intentional and was a path to network effects; the more people who joined Calendly, the more they unlocked these additional benefits.

The idea of critical mass is also crucial to engineering network effects. It gives you a sense of when to invest in it earnestly. For example, Microsoft Word reached critical mass by being distributed on almost every PC for years. As a result, most people started to demand files be sent to them in the Word format. And then the more people did this, the more others who needed word processing demanded to have Word installed on their PCs, which is a network effect. However, investing in a specific file format before the point of critical mass would have been suboptimal or would not have yielded the intended benefit. In fact, in the early days, Microsoft focused on importing and interoperating with the WordPerfect file format, which was dominant then.

Software creators should always have the engineering of network effects in mind. Build the foundation as early as possible by considering both direct and indirect ways to create more value for customers as the network grows.

Access to Capital

A key source of competitive advantage is maintaining access to capital. An effective annihilator of a software company, especially in the early stages when it's not profitable, is running out of cash.

Capital access covers a lot of tactics and investments across the growth lifecycle of a software company. Early on when a company is private and likely unprofitable, it should focus on maintaining adequate working capital. This is usually in the form of any kind of venture capital or bootstrapping. As a company grows it can focus on profitability or growth. A growth focus implies either more venture capital injection or the ability to maintain access to credit.

Ultimately a company can maintain access to capital by orchestrating an initial public offering in the stock market and working hard to maintain a healthy stock price. This unlocks secondary access to capital, like favorable credit and corporate bond issuing.

When comparing similar companies, the one with greater access to capital has significant leverage because they can fund more speculative research and development or use it to increase their market distribution or broker advantageous partnerships. Access to capital can also fund mergers and acquisitions (M&A) that can grow the rival's position in the market, in addition to revenues.

Access to capital can be boiled down to a few maxims:

First, don't ever run out of money. Many potentially great companies have been slayed by not being well capitalized. Oji experienced this quite directly as a teenager when his dad—an entrepreneur—could not secure working capital for a small, highly valuable manufacturing venture. Indeed, many promising companies and funds were slayed by the venture capital crunch of 2022–2023.

Second, diversify your capital access in a way that allows emergency spend on opportunistic investments or gives you the ability to weather unexpected adversity. The 2020–2022 global pandemic leveled many promising software businesses because of the drop off in demand for certain kinds of software products. Various financial crises can have the same effect as well.

Ezinne has been really lucky to work with some really experienced CFO's from Bazaarvoice, Procore, and WP Engine and got to see how great operations teams managed their financials as well as maintained constant communication with capital sources. These served the companies well, such that there was hardly ever an investment opportunity that they could not take advantage of.

OpenAI has demonstrated remarkable versatility in accessing capital across different stages, starting as a non-profit with $1 billion in pledged funding, then evolving to a hybrid structure that attracted a $1 billion investment from Microsoft in 2019, followed by a $10 billion extension in 2023. They've also successfully diversified their capital sources through product revenue, reaching over $3.4 billion in annualized revenue from their API and ChatGPT subscriptions by June 2024. This diverse access to capital allows OpenAI to fund expensive AI research and development while maintaining their technological lead, even as compute costs and talent expenses continue to rise in the competitive AI landscape.

Regulation

Favorable regulation can provide a really strong advantage in any market relative to the competition. If regulation spurs smooth and plentiful demand vis-á-vis your competition, or somehow your company's ability to take advantage of regulation is higher than the competition, then this creates significant competitive advantage.

Very popular examples of this abound. Healthcare tech in the US has been significantly boosted by the Affordable Care Act and its incentives for electronic medical records and other healthcare automation. Cleantech is having a huge surge at the moment because of the Inflation Reduction Act of 2022. These are only US-centric examples; the same principle applies in every economy in the world. In Nigeria, legislation boosted cell phone companies, digital banking, and ID card services across the nation, enriching companies that participated and enabled these government-engineered outcomes.

Regulation is such a significant driver of advantage that many scaled software companies maintain a permanent lobbying presence in every state capital of note to their business, in order to influence legislation in their favor. This is because, in the same manner that it can be a boost, hostile legislation can be a significant headwind to existing businesses. For example, Section 230 of the Communications Decency Act of 1996 in the United States absolves software content platforms that are user-generated from the repercussions of the speech from customers and contributors on their platform even as they continue to monetize those platforms.

A repeal of this would destroy billions in value from social networks and other publishing platforms, therefore a lot of lobbying goes into preserving it.

Another example of regulatory heft is Intuit, which has turned tax and accounting lobbying into a formidable competitive advantage. Their decades-long relationships with tax authorities, and deep integration with government systems, have shielded them from competition for decades, such as Microsoft's failed attempt to overtake them in the 1990s. This regulatory expertise has helped Intuit maintain its dominance in tax preparation software despite numerous well-funded competitors.

LOOK FOR YOUR EDGE

After a successful startup phase, the objective of a software company is to create enduring profitability. The only way to do this is to develop and stack competitive advantages. Start with the dimensions firmly in your control: your IP, brand, and market presence. Become indispensable to your customer and deliver on your promises. Then, as you grow, navigate toward new advantages: economies of scale and scope, vertical and horizontal integration, and network effects. Use your momentum to gain an edge in your access to capital, and when you reach the mountaintop, maintain your position by pressing on the levers of regulation.

This is the playbook run by every world-changing software company, and while the tactics change, the strategy does not. Like the last chapter, each of these dimensions of competitive advantage could be an entire book. We've just covered enough for founders, product builders, and tech company execs to wet their feet and have some way points as they think of investment directions. We make recommendations for further study in the Pro Edition and in our ProductMind Community.

The previous two chapters explored key growth channels and competitive advantages. Competitive advantage is at the heart of any direction-setting system. All growth strategies should ideally be chosen to confer competitive advantage now or later. While your executive team should regularly evaluate and strengthen these growth paths that create advantages, these ideas aren't just for the C-suite. Every member of your Shipyard needs to understand your growth lanes and potential competitive strengths as they spot opportunities and build solutions. The sharp problems you choose to solve, the customers you target, and the way you build your solutions—all of these decisions should reinforce and deepen your competitive advantages. The most successful product-led companies don't just stumble into strong market positions; they systematically identify, build, and defend their competitive advantages at every level of the organization. In the next two chapters, we will discuss how to turn strategy into action and strong execution.

FOURTEEN

Launching the Ship: Systems of Execution

Often when we talk about systems of execution, people immediately think about engineering execution—how to navigate concepts like backlogs, epics and stories, etc. There are even more abstracted development systems like the Shape Up system invented by 37signals (we like Shape Up a lot! Check it out). Your product system of execution definitely contains the engineering execution pieces above. However, in our experience, the gap between strategy and engineering execution is where most product teams stumble. We've seen it countless times—a brilliant strategy gets lost in translation as teams move from direction-setting and planning to building. The result? Stalled progress, misaligned efforts, and frustrated teams.

Why does this happen? Through our years leading product teams, we've identified five core execution challenges:

1. Teams lose sight of the connection between direction and priorities and their projects. When squads can't connect their daily work to the broader mission, they drift off course.

2. Cross-functional teams stop collaborating effectively. Different departments fall out of sync, leading to wasted effort and missed opportunities.

3. Teams get stuck in analysis paralysis. The quest for perfect information paralyzes decision-making.

4. Project management becomes unfocused. Without clear processes, deadlines slip and deliverables remain incomplete.

5. Communication breaks down. When information stops flowing freely, trust erodes and alignment suffers.

While there are excellent resources on engineering execution—the mechanics of agile development and similar frameworks—we won't retread that ground here. Instead, this chapter focuses on something more fundamental: *how to align all Shipyard efforts with your product strategy*. We will just assume that once you have your priorities and projects correctly defined and pointed in the same direction as high-level strategy, that you can turn them into engineering tickets and the Shipyard can get to work.

This part of the system of execution, the focus of this chapter, dwells on the intentionally designed practices, artifacts, rituals, and processes that keep your Shipyard building products that delight customers and drive profit. It's how to take lofty strategy, full of carefully crafted growth levers and competitive advantage points, and turn it into actionable things to build. This gives your entire organization confidence that the backlog, epics, and stories are focused and aligned with the mission. You've done the hard work of developing your strategy. Now comes the crucial part: getting your entire product organization to execute on it effectively. This requires two key elements.

First, your strategy must be widely known and accessible. Leaders need a system to communicate direction frequently and clearly. Anyone in your Shipyard should be able to find and understand the strategy in an hour of focused absorption.. We recommend starting every important meeting with a two-minute reminder of the product team's mission and priorities. Many people will roll their eyes at this, but it's a small, useful tool to keep everyone aligned with your strategic direction.

Second, every team or company has to update its strategy at some point. As time progresses, objectives are attained (or not), and the market changes; at any rate, any strategic blueprint will have a natural half-life. We recommend that product strategy be updated in six to eighteen-month cycles unless the company is in a new market that has many unexpected disruption. The actual interval is usually governed by the pace of change in and outside the company.

It's useful to have a systematic way to update strategy. Smart product leaders craft a system for this with clear principles and methods to use in changing direction from what it was to a new direction, when necessary.

These two parts of operationalizing strategy are fundamental skills of product leaders, so let's explore both in more detail.

DOCUMENTING STRATEGY

We highly recommend that the strategy be a live doc or presentation. Product leaders should publish strategy docs, explainers, and other artifacts in a well-known location. These artifacts should have a comment function so that others can react to it.

It's important that the team asks questions about strategy and has an opportunity to point out flaws and improve it. In turn, these documents should be updated from time to time to reflect external input from real-time conditions.

COMMUNICATING PRODUCT STRATEGY

Now you have a strategy. What is the most concise way to share it? In this section, we discuss some of the major things we believe product leaders need to communicate strategy effectively to a team that can execute it.

There are many frameworks for accomplishing this, but our go-to is the VMSOO-P (Vision, Mission, Strategies, Opportunity Costs, Objectives/Goals, and Priorities). We'll get into what each component means and how to articulate it properly for your strategy document.

The important thing is that we advise leaders to create critical assets and rituals that clearly speak to the items in the framework for the whole company. Our ideal is actually three types of assets and a few rituals.

Assets:

- The primary strategy document (Word, Docs, or Coda).

- The presentation deck that takes from the document and highlights key points for those who prefer a visual medium. It also forces teams to boil things down to their essence.

- A video talking through the presentation and providing context, recorded and used as part of onboarding for new employees.

Rituals:

- Roadshow for the kickoff of any new strategy.

- Periodic updates to discuss progress or any change in direction.

Vision

A vision is nirvana. It's the dream—a team's true north. An articulation of how the world has changed if your company is wildly successful beyond your dreams. It's better if it's evocative and specific. The point is to inspire wonder and create a shared purpose throughout the company. A good vision is opinionated about the future, not just the present, and should inspire the team to action. The scale of the vision should be just on the edge of attainable, but only with great and coordinated effort.

Famously, Microsoft articulated a vision of "a desktop in every home." They did this when the idea was still very far-fetched for most people. PCs had just been invented and the main model of computing was the mainframe, like Digital Equipment Corporation's VAX system. Now, at least in the West, this dream has been achieved many times over and has gone far beyond the original vision (supercomputers in every pocket). In our own home, there are over one hundred different sensors and computing devices connected to the internet, and only about twenty would be considered desktop class.

Mission (Your North Star)

The mission is a specific major objective from which the rest of the VMSOO-P flows. We introduced the concept of a North Star in Chapter 7—mission and North Star are synonymous.

Your mission is the most immediate achievement that can make the vision possible in a reasonable amount of time, ideally within one or two business cycles, or three to ten years. It should be measurable, achievable, and inspirational. It should be brief, easy to remember, and ideally proves to be uniquely identifiable to the product or company.

An example of a strong, compelling mission was T-Mobile's mandate to become a top two mobile carrier in the United States in the early 2010s. From this mission led T-Mobile's commitment to the Uncarrier brand and strategic partnerships with Starbucks and Amtrak.

Visions and missions are often used interchangeably. To our mind, the key difference is that a vision articulates a future aspirational state of the world, while mission is a specific, verifiable outcome. You should be able to tell whether a mission was achieved with incredible clarity.

Strategies (Your Levers)

Strategies—we also call them levers, or the specific moves and investments that a team makes to win the market—express how a Shipyard team navigates its mission and its competitive landscape to achieve its objectives. The best strategies outline what to do and also what NOT to do (the opportunity costs). Calendly's strategy was built on its durable virality, engineered network effects, and its horizontal integration with all calendaring apps. The company chose to not pursue strategies that put them in direct competition with Microsoft and Google, such as creating a developer ecosystem, calendar app, or email product.

Your strategy can be expressed as the levers that help your Shipyard achieve competitive advantage and growth—remember Chapter 13? Typeform developed specific metrics (which are not public) to quantify *profitable growth*. They pulled several strategic levers to achieve this mission, including *reach levers* like increasing the activation rate and *value levers* like improving the core application with the use of large language models and other AI techniques.

When you have a moment, take a look back at to Chapter 12 (Growth Levers) and Chapter 13 (Competitive Advantage) to review your strategic options again.

Opportunity Cost

It's important to articulate strategies that will NOT be employed by the team or company and it is important to be both reasonable and specific. For example, you could grow a company internationally, but it may not be the right time because the infrastructure to do so is not in place yet, and therefore a strategy based on this in a given period may be ineffective.

When developing strategy documents, Ezinne addresses opportunity costs by weighing the pros and cons of different strategies. This shows that a) she has considered options other than the recommended or chosen approach, and b) she allows her colleagues to conduct their own cost/benefit analysis and make suggestions. Being clear about what will be left on the table or not worked on with the chosen strategy is critical for the executive team and the organization. It serves as a record of our consideration and it helps keep us accountable, as there will always be pressure to do "that really good idea" or to follow the path not chosen. Putting down on paper that "for right now, this is not our focus and we understand what that means" can be galvanizing as long as teams are aligned, which must be done in the strategy setting process. But, if or when we do have to pivot and change course, we can do so by acknowledging any new insight or market moves that necessitates considering a path we had initially documented as forgone or an opportunity cost.

Objectives and Goals

There are dozens of potential paths a company can take to achieve their vision, which is why you need a full list of levers before you cull them. To fully understand which strategic levers may be most effective, quantify your strategy with achievable objectives. Note: Chapter 7 goes a lot deeper into goal-setting.

Objectives give the team a consistent scorecard of success that is universal and verifiable. In the absence of clear objectives, teams often create their own measures of success that may be divergent to the objectives of the team or company and thus may not achieve its mission.

To continue the Typeform example, the team identified several objectives and goals that, if achieved, would result in profitable growth for the company:

Objective	Goal
Increase customer activation	Achieve XX% growth of customers activated
Maximize form submissions	Achieve XX million form submissions in 2024
Stabilize customer retention	Achieve 6 months retention of 30% of activated customers
Expand revenue	Achieve +$5/customer over last year per customer revenues

Some of these goals are not public information.

How did they arrive at these objectives? You can trace the logic steps up to the strategic levers, then the mission, and finally, the vision. Anyone on your team should be able to see how their specific objectives level up to support the overarching company direction. Objectives should not be pulled from thin air, and product leaders should be sure all the objectives *combined* will result in a successful mission. Remember that careful aiming is half the battle of achieving the mission.

Priorities

With your objectives set, you can now drill down to specific priorities. Priorities are ranked lists of projects that are selected and designed to help the company realize its objectives as fast and comprehensively as possible.

Careful selection is important; the essential question is, if we could only do one thing, what would it be? It's rarely one thing, but this is still the most pragmatic approach to focus your team's efforts.

How you rank your projects is critical if you want to move with any speed. Which of these things will have the biggest impact? The goal should always be to work on the next most valuable initiative.

Typeform's customer activation objective resulted in priorities that looked something like this:

1. Streamline the onboarding process.
2. Implement an AI-driven personalization system.
3. Develop a gamified tutorial for key features.
4. Create an email nurture campaign for new sign-ups.

Each of these projects is ranked on how important they are in their relative contribution to the objective of increasing activation rates.

Priorities are where the rubber meets the road. It's how you turn your strategic vision into products and services. Therefore, it's worth our time to dive deeper into the art and science of setting priorities for your teams.

SETTING AND MANAGING SHIPYARD PRIORITIES: WHAT TO EXECUTE AND WHEN

Now comes the challenging part: deciding which projects or ideas to tackle first and how to allocate your resources. Here's a step-by-step process we've found effective.

1. List all potential projects

Start by listing out all the potential projects you could pursue. Don't filter at this stage—get everything on the table. Remember, this can be done for a specific squad/team or even the entire Shipyard (usually part of an annual planning cycle at that scale). For larger organizations especially, this is good time to take stock of the projects, get a sense of the different categories they fall in, and determine if you want to first assess impact/effort for each category independently and then rank together. Your approach here will determine if and how to calibrate on impact, which is a critical overarching ethos that is required in priority setting within an organization.

2. Assess impact and effort

For each project, assess its potential impact on your goals and the effort required to implement it. We often use a simple 1–5 scale for each dimension, where high impact/high effort is 5 and low impact/low effort is 1. There are so many models for prioritizing out there with additional dimensions to consider for the projects. What we've learned is that if the teams can align on impact and effort and get a solid list based on this, alignment gets simpler because we are now speaking generally in the same unit of measure.

3. Calculate the impact-effort ratio

Divide the impact score by the effort score. This gives you a rough measure of the "bang for your buck" for each project. For example:

Project	Impact	Effort	I-E Ratio
Project A	3	4	0.75
Project B	**5**	**2**	**2.5**
Project C	4	4	1

4. Consider strategic importance

Some projects might have outsized strategic importance beyond their immediate impact. Others may drive impact but also be enablers of other projects. Make sure to factor this in—the sum of impact should be how much it impacts the current goals (immediate impact) and how much it impacts longer-term strategy. For example, Apple chose to add AI chips to the iPhone back in 2017. They were inactive but will now power the early versions of Apple Intelligence in earlier phones. Some teams choose to add a separate strategic impact column to their analysis to reflect this.

5. Evaluate dependencies

Consider any dependencies between priorities. Sometimes, a lower-impact item needs to be done first to *enable* a higher-impact one with difficult-to-navigate dependencies. At WP Engine, Ezinne's team would actually take a break after step 4 to have the engineering teams stare at the list and extract the core dependencies they could see and then help articulate a recommended sequence. This would factor high into Ezinne's consideration when it was time for the tough calls.

6. Make the tough calls

Using all this information, make the tough decisions about which priorities to pursue and in what order. Go into those sessions reminding everyone involved that: "Today we will be making some hard decisions, but that's our job, to leverage our knowledge and insights to make tough calls. There's human judgement required." (If there wasn't, we'd be able to monetize an Impact/Effort mega-spreadsheet for all software companies and become incredibly wealthy.) Making tough calls has a way of galvanizing teams and reminding them about the art of their jobs as product managers, leaders, and executives.

Pro Edition: Strategy Templates

We share our VMSOO-P template and prioritization worksheet in the Pro Edition of *Building Rocketships*. These docs will help you set direction and keep your team aligned. Get them now: productmind.co/brpro.

GLOBAL PRIORITIES: ALIGNING ACROSS THE ORGANIZATION

While individual teams may have their own priorities, it's crucial to establish a set of global priorities that span across the entire organization. These global priorities help prevent siloed thinking and ensure that all teams are rowing in the same direction.

Global Priorities	Team Priorities	Team Priorities	Team Priorities
1	3	1	2
2	7	4	6
3	8	5	9
4	10		
5			
6			
7			
8			
9			
10			

Create a master list that ranks all priorities from most to least important, regardless of which team or strategy they belong to. This global prioritization is vital for two reasons: First, it prevents teams from working on lower-priority items (say, priority #9) when higher-priority items (like #6 and #7) haven't been addressed yet. Second, it facilitates resource reallocation when necessary. If a team finishes their assigned priorities, they can easily see where their efforts are most needed next by consulting the global priority list.

BALANCING SHORT-TERM AND LONG-TERM PRIORITIES

One of the trickiest aspects of priority setting is balancing short-term wins with long-term strategic investments. It's tempting to focus solely on quick wins that show immediate impact. But this can be a recipe for long-term stagnation.

We recommend using a portfolio approach to calibrate the efforts of the Shipyard:

- 70% on short- to medium-term priorities that drive immediate business results
- 20% on longer-term strategic bets that could be game changers
- 10% on exploratory work to uncover new opportunities

This balance ensures you're delivering value now while also investing in your future.

Communicating Priorities

Setting priorities is only half the battle. You also need to communicate them effectively to your entire Shipyard. Here are some best practices we've honed from multiple companies at different stages, in different industries:

1. Be crystal-clear: There should be no ambiguity about what the priorities are and why they matter.

2. Provide context: Explain how each priority ties back to broader goals and strategy. Quantify it and make it an almost mathematical relationship. In certain cases, we recommend writing an equation.

3. Be transparent: Share the reasoning behind your prioritization decisions.

4. Reinforce often: Don't just communicate priorities once. Reinforce them in team meetings, one-on-ones, and company-wide communications.

In addition to the core VMSOO-P document, product leaders should pursue various graphic and visual artifacts that convey the direction of the product team to different audiences. What is needed depends on your audience, the company, and its size.

When Microsoft was about 75,000 employees, the company would rent a very large stadium every year, bus all employees to it, and use giant screens to inculcate its direction for the next year across all its many teams and employees. This was a big production and spectacle. That likely cost millions of dollars, but it was deemed worth it to create shared direction and visibility of plans for everyone at the company.

This practice generally achieved its goals, even at Microsoft's scale. If a giant company can prioritize communicating vision and direction twenty to thirty years into its existence, so can a young startup or still-growing company.

Communication Opportunity Costs: The Power of Saying No

Perhaps the most crucial skill in priority setting is the ability to say no. Every YES to one project is implicitly saying NO to countless others. Embrace this. A clear no is often more valuable than a hesitant yes.

When saying no, be respectful but firm. Explain your reasoning, tie it back to your prioritization framework, and if appropriate, suggest when it might be reconsidered in the future.

Supplemental Communication Materials

Here are some items we have seen be successful as supplemental communication material:

1. **Strategy Document**
   This is a longer version of the VMSOO-P strategy in document format. It should cover both the state of the world, the state of the

company and the product, the strategic options and the strategic choices made, and why they are compelling. It's a brief discourse of the strategy that is suitable for everyone in the company beyond the brevity of the VMSOO-P.

2. **Strategy Movie Script**

This is an optional narrative device that imagines what the product and customers will be like when the priorities have been achieved. It creates a compelling story of the end state of the strategy that can inspire the team on what they're shooting for. It's best if this communication method is also visual. Usually this document tells the story of the end state of the mission/goal if achieved, not the vision. This is a pragmatic take because the vision is meant to be slightly unattainable, so the movie should feel very attainable. This is a very practical way to involve the design team in strategy.

3. **Strategy Presentation**

This is largely a deck version of the strategy document. In many cases, using a long-form document is not appropriate and you need to tell the story sequentially in deck format. This format is actually the most common in many organizations (we heavily favor long-form writing, but it's not common, and relevant audiences love more visual fodder). We try to convince product leaders to use documents (see above) as well, since they can contain more detail for your product leaders.

4. **High-Level Roadmap, or Press Releases**

A strategy is best accompanied by a very specific set of problems that it will solve in a given time frame, i.e., a roadmap. If a VMSOO-P spans the next three years of product development, it should at least envision the key problems that will be solved within, say, a twenty-four-month timeframe and key solutions that you want to offer to solve those problems. A high-level roadmap is a good document to negotiate with CEOs and boards that want specifics. They should not be 100% of what is to be shipped, just the top things and with enough of a planning buffer to ensure success.

5. **Objectives/Goals Breakdown**
   Goals should be specific for the Shipyard and for each team. And they must be connected. If possible, show the math of how the North Star goal is mathematically related to the few sub-goals that support it.

For example, at Typeform, the leading indicator is form submissions. To arrive at this, the team needs to influence the number of customers acquired and average submissions per form:

$$[\text{submissions} = \text{number of customers acquired and retained} + \text{avg submissions per customer}]$$

This leads us to influencing activation rate (how many customers learn how to use the product quickly), conversion rate (how many customers make us their tool of choose for their job to be done), retention rate (how many customers stick around and keep paying), and the average number of use cases per customer. In this manner it's easy to illustrate the goals:

- North Star = form submissions

- Supporting goals = activation rate, conversion rate, retention rate, and average number of use cases per customer

Furthermore, each supporting goal is best assigned to a main group or team. That way product leaders are confident there is solid goal ownership and clear responsibility lines for key goals. Other teams can contribute to a goal, but a major owner is key to holding teams and individuals accountable.

In practice, VMSOO-P can exist at multiple levels of the team—at the Shipyard level, at the group level (multiple teams with a same or similar mission), and at the squad level. We recommend it exists at the Shipyard team level, i.e. encompassing the entire set of teams that build the product. Thus, you can have as many VMSOO-Ps as there are products in a software company.

However, it's also useful to have a MSOO-P[22] at the squad level. This helps squads focus on the key activities and projects that they bet will accomplish their goals and ladders up to the product goals. Often a specific mission within the product group mission may be warranted. For example, a squad may want to build the best AI platform that serves the product its group is building as its own specific mission to contribute to the larger product objective.

Resource Allocation: Giving Your Priorities Life

Once you've set your priorities, it's time to allocate people and resources. Think of this as placing bets in a high-stakes game. You have a limited number of chips (your squads), and you need to decide how to distribute them across your priorities for the best possible outcome.

Generally a Shipyard leader like the CPO will allocate one or more squads to initiatives. Then product directors and principal PMs will match priorities to the effort or execution sequence of those squads.

Some initiatives might warrant a full squad or more, while others might need just a fraction of a team's time. Be prepared to make bold moves—sometimes, the right call is to go all-in on a single, high-impact priority. Making the right bet is the job of product leaders at all levels.

Remember, resource allocation isn't just about headcount. Consider all your resources: time, budget, tools, and even management and coordination overhead.

22　There is really no need to repeat a vision-level statement at the resolution of a squad/team.

It is hard to stress how precious your team's collective leadership attention is. This is why we recommend making fewer high-impact bets vs. many medium-impact ones. It's important to know just how much management time certain priorities will need and track this so you know when your leadership attention budget is exceeded. We have all seen projects that survive only because of the herculean efforts of managers and high performers—those who "jump in" to get things across the finish line. Many projects "mysteriously" fail simply because leadership is not paying adequate attention.

Continuous Evaluation and Adjustment

Objectives and priorities selection isn't a once-and-done exercise. It's an ongoing process that requires constant evaluation and adjustment. We recommend reviewing priorities quarterly with check-ins that make sense for your organization, like every six weeks. Smaller companies may need more frequent check-ins. We've seen in some larger companies the cost to change or modify priority, and in those cases, accept that some companies can only truly adjust priorities every two quarters due to the toil of moving people and resources around too often. If this is the case for you, it's worth thinking about the size of the projects in your priority list as well as what investments can be made to create more nimbleness or agility.

For these evaluation sessions and check-ins, product leaders should ask themselves:

- Are we making progress on our key priorities?

- Have market conditions or customer needs changed in a way that should shift our priorities?

- Are there new opportunities or threats we need to account for?

Be prepared to make changes based on new information or changing circumstances. Agility in priority setting can be a significant competitive advantage, just as institutional inertia can be ruinous.

Setting priorities is one of the most important—and challenging—responsibilities of a product leader. It requires a combination of strategic thinking, analytical rigor, and decisive action. Get it right, and you'll focus your team's efforts where they can have the most impact, driving your product and business forward.

Remember, in the Shipyard of product development, your priorities are the navigational system that steer you where you want to go. Choose them wisely, communicate them clearly, and be ready to adjust course as needed. Your team's productivity, your product's success, and ultimately, your company's future depend on it.

What priorities will you set to guide your Shipyard? And more importantly, what will you say no to, to make those priorities a reality?

Closing the Loop on the Results of Priorities and Goals

You've set clear objectives and priorities. You're evaluating and making adjustments. What remains? Continuously closing the loop. In the next chapter we talk about alignment. However, it's important to note here that a sizable Shipyard looks like chaos from the outside. Customer teams, business teams, and executives *cannot* keep track of all the moving parts.

Smart Shipyard leaders, especially the CPO, have a responsibility to not only set direction, distill objectives and priorities, and measure attainment of these plans, but also communicate how all this is tracking to expectations.

Specifically, as priorities change, you need to communicate them (usually slightly different variations based on audience) to the Shipyard, execs, and the board. As the Shipyard accomplishes its goals (or not), you should send progress indicators that allow alignment to be accomplished.

THE BRIDGE TO AGILE

In our experience, many product teams face execution challenges during the transition from strategy to actual development and implementation. It's in this handoff that execution often stalls or in connecting all the execution activity to progress in achieving said strategy. Issues like unclear priorities, misaligned teams, poor communication, and over-analysis can lead to delays and roadblocks that derail progress.

The top culprits include:

- Lack of clear direction, alignment, and prioritization
- Cross-functional miscommunication
- Analysis paralysis
- Poor project and time management
- Insufficient communication

We chose to focus this chapter on those impediments to execution—versus the actual systems of software development, like agile—that address effective engineering implementation. Your chosen agile development process should feed your execution system, and there are many excellent resources that discuss agile in detail.

Let's zero in on the systems that bridge strategy and execution, as well as help reconnect back to why the communication of progress matters too. When you have clear and aligned priorities that are well communicated, those then can be broken down into the right development unit, whether they be epics-of-epics, epics, stories, or tasks. These should all sound familiar to any member of a Shipyard. The important thing, then, is to ensure that the output of the agile process gets repackaged up to the rest of the organization in a way that matters to them, or a way that helps them understand how all the activity and effort is moving our strategy forward.

FIFTEEN

Changing Direction: Modifying Strategy & Alignment

Given how quickly the market can move, technology companies often need to change direction. Startups do this even more often because they effectively have no set strategy until they achieve product-market fit. Larger-scale companies often need to see a strategy executed through half a business cycle in order to evaluate if it's working well and whether direction change is needed.

However, these rules of thumb only really work when the waters surrounding the business are fairly calm. Everyone has heard of the pivot. Sometimes startups will reach a major blocker in finding PMF or are forced to change course to respond to a lack of money, pressure from investors, and more. Even larger companies may be forced to respond to sudden market changes. The early 2020s offered many of these sudden market changes: The first was the Covid pandemic circa 2020–2021. The second was the market decline circa 2022–2023, challenging capital access, thus necessitating many companies adopting profitable growth as an ambition. The third was the emergence of easily accessible, API-driven generative AI in 2022.

Any of these would be challenging in and of itself to technology companies, but all three together became a perfect storm of disruption and an existential event across the economy.

So how does one change direction? And why?

In short, modifying direction is an effort to react and right-fit the situation that a technology company or startup finds itself in, relative to its activities and goals. This is often necessary because not adjusting will lead to lack of growth, falling behind the competition, or even company death. The history of the tech industry is a graveyard of companies that did not react quickly or correctly to change. It's worth illustrating a couple before outlining some of the techniques for enacting more sure-footed change. So let's discuss two quick examples.

Netscape vs. Microsoft: A Successful Pivot

In 1994, Netscape (started by Marc Andreessen and others) launched their internet browser, the first major commercial browser offered to consumers[23]. It was an obvious challenge to the Microsoft Windows Operating System. Because of the emergence of the World Wide Web, Netscape threatened to become the center of the software universe vs. the native operating system. The browser was already replacing some of the functionality of the OS, like using hypertext as a file retrieval system over the internet instead of a local file system or a local area network. The browser could also render information and media and do so across operating systems, making it an attractive platform to build on instead of Windows.

At the time, Microsoft was executing a five-year strategy built on unifying its OS versions into a single system. It was working on a 64-bit version of Windows that could work on both consumer desktop as well as workstations. When Netscape was released, Microsoft leadership decided this threat had to be taken seriously and orchestrated a hard pivot in strategy.

In quick succession, Microsoft licensed the Mosaic browser and with some small modifications offered it as Microsoft's Internet Explorer. In 1995, it embedded the

---

23   It was not the first browser, however. Andreessen, along with Eric Bina, released the Mosaic browser in 1993 under the auspices of his academic work at the University of Illinois at Urbana-Champaign.

technology deep into its Windows 95 operating system, blurring the difference between the OS and the browser. It then began to make custom extensions to web browsing standards that only worked on Windows. It wanted to tilt developer attention to favor its Internet Explorer browser over Netscape's. Because Microsoft had a big distribution advantage through its dominant operating system, it initially won out. Netscape was a much smaller company in a new industry dominated by Microsoft. By the time Netscape's antitrust complaint to the US government was settled, it was too late to save the company from the fierce and surprisingly swift response of software technology's "900-pound gorilla."

By 1999, Netscape was struggling to keep its doors open and eventually sold to America Online (AOL).

Microsoft changed direction very quickly to confront competition and its timely, aggressive actions kept its Windows OS the dominant operating system for personal computers all over the world till this day. Microsoft later lost many battles related to the browser technology and future battles for dominance, e.g., in mobile operating systems and cloud technology; but none of these threatened its core business as severely as Netscape had, and its actions remain decisive to its longevity.

Nokia vs. Apple: A Failed Pivot

In 2007, Steve Jobs unveiled a small mini-computer that was also a phone. This device had a competent touchscreen response not seen before and very quickly (subsequently) became a full app development environment.

The state of the art just before this was threefold: Nokia's excellent phones, which ran the Symbian operating system; Research In Motion (RIM) Blackberry; and Microsoft's Windows Mobile, which was supported by many PC hardware OEMs. From a small start on one US carrier, the iPhone began to outsell all these competitors. Nokia had the most to lose because it had about 50% of the cell phone market.

The iPhone was sold at a premium and was immediately profitable, eventually growing to take the lion's share of the profits in the smartphone market which it retains, circa 2025. For up to three years after the iPhone was launched, Nokia continued to ship more cell phones per year than the iPhone, but its profitability dropped significantly. By 2011, its CEO, Stephen Elop, understood that the company's position was untenable at its scale. Nokia had already fired thousands of employees, but it still could not compete effectively because its software (and its attempts to replace Symbian 6) could not match iOS's sophistication.

Nokia chose to adopt a new OS but struggled to decide which one. The most viable OS's were Android and a new operating system by Microsoft called Windows Phone. Android seemed like a nonstarter because Nokia's conventional rivals like Samsung had already adopted it. Choosing Android would erase Nokia's pre-iOS software advantage and turn the market into a pure manufacturing slugfest with the likes of Motorola and Sony Ericsson. By choosing Windows Phone, Nokia gained some differentiation with a partner that would not compete with it and with which it could have a more exclusive agreement.

Ultimately all these interventions failed because consumer demand for the iPhone and its ecosystem was high, which maintained its profitability. Nokia could not create a differentiated offering in the midst of competition from strong new entrants like Samsung and others. To survive on razor-thin margins and lower volume, Nokia had to shrink and ceased to exist in the shape that it was before the introduction of the iPhone. Its hardware business was sold to Microsoft but with only limited rights to the brand. Eventually, the brand re-entered the smartphone market selling mostly hardware devices created by various hardware partners.

In retrospect, the three years after the introduction of the iPhone proved to be a crucial time to make changes to its strategy. Instead, Nokia did not fully appreciate the extent of change that had happened in its market. The pivot came too late to prevent decline. Market shifts can be catastrophic for businesses with a hardware element if they misjudge change.

The margin of error in software-hardware businesses is much smaller than in a software-only business. It's no surprise that the advent of the iPhone killed many companies—RIM, Nokia, and even Microsoft's Mobile OS. However, it also gave room for new, more competent competitors to rise, like the Android OS and hardware manufacturers like Samsung, Huawei, and others.

HOW TO CHANGE DIRECTION AND SET NEW STRATEGY

Changing direction is a matter of mapping the current plan with reality. The job of direction change is the work of founders or company leadership teams. As a company scales, change must be managed as a team sport to ensure that it can quickly percolate deeply in an organization. Here is a small set of suggested steps that can work at almost any scale.

Gather a team responsible for updating the direction

Choose experienced and credible leaders from various departments who possess the necessary skills and knowledge to gather and analyze crucial information. These individuals should have a track record of making sound decisions and executing strategies effectively. Consider involving leaders from product development, marketing, sales, and finance to ensure a well-rounded perspective.

The larger the company, the more people will want to be involved in the direction change. Fight against the urge to make the team bigger upon request. Instead, set up a level of involvement: maintain a small planning team (five to seven leaders max), then create a second ring of leaders who are kept in the loop and asked for feedback.

Steep them in a shared reference of the current trajectory

The people responsible for direction change, or more generally, setting strategy, should be steeped in what the current strategy is—its objectives, its reason for being, and the circumstances that created its necessity. This kind of historical understanding prevents a certain kind of blindness that afflicts companies—as talent circulates (comes and leaves), people forget things that have worked or haven't worked and people repeat mistakes. An organization that doesn't learn effectively as employees cycle in and out is a dangerous place to work in.

Next, they must be conversant with how the strategy has performed. What went right, what went wrong? Why? This matching up of the strategic bets with reality is a natural retrospective that can help with calibrating future bets and approaches in order to avoid past mistakes.

Usually, sessions that examine the current strategy and its performance are a critical foundation to any strategy-shaping activities.

Either way, a thorough understanding of the internal and external environmental challenges is key. This is akin to the customer discovery needed for building an impactful roadmap. Strategists must understand the historical environment and the current one to sufficient depth in order to point in directions that can result in material growth of the company.

Account for changing company circumstances

The issues perturbing the market have to be well analyzed. In the case of Nokia, what was needed was an evaluation of how much Apple's success with the iPhone would change the market, and not in Nokia's favor. They needed to play forward a pre-mortem view of the world. But it's likely that all Nokia did was play forward a world in which the iPhone failed and they fought back and won with all their manufacturing prowess (all in-house in the US and Finland/Europe, not outsourced to China).

A better process was to take the entirety of the situation in hand and work out both rosy version of the future and a not-so-victorious version of the future, and then start to put countermeasures in place to deal with the negative possibilities and eventualities.

In the same way, when Oji was at Atlassian, the company was preparing to assault the work chat market—which they had pioneered with HipChat—but that Slack had taken by storm. By building Atlassian Stride,[24] Atlassian was striving to bring a next-generation perspective to the market and win. However, things changed when Microsoft entered the market with Microsoft Teams. Atlassian was prepared for guerilla warfare with Slack, another venture-funded company. But it was not prepared for a two-front shooting war with two companies with gigantic arsenals. Microsoft especially had a fearsome distribution advantage with Microsoft Office 365. As part of its direction change, Atlassian decided to get out of the market and sold its IP and code to Slack.

This is a stellar example of direction change that comes from updating information about a company's changed circumstance, and the market environment. Atlassian wisely bowed out of the battle to win the war.

Decide how to decide

Setting the direction of a product team or company can be hard. Usually if a company has any scale, it will have many options, but choices must always be made about the best use of resources. The planning team or even those above them (the leadership team) must lay out a set of rules and conditions to help them pick a

---

24   Reynolds, Annelise, "Atlassian Redefines Team Communication With Stride," Atlassian, September 7, 2017, https://www.atlassian.com/company/news/press-releases/atlassian-redefines-team-communication-with-stride.

path from the options available. The key questions are: What are we optimizing for? What criteria has to be met with a good plan?

Often the simplest criteria is maximizing growth in terms of customers and revenues. So whichever option achieves this maximization is selected. But there may be other options. In other economic climates it might be maximizing profitability at the expense of growth (this is particularly relevant in the climate for software technology companies circa 2024 when this is being written). Other options are maximizing cash on hand or other defensive positions in anticipation of tough economic times. It could be successfully positioning the company for a sale or an initial public offering and thus nailing the required metrics for such an event would be the prime focus.

At any rate, how to decide is critical because otherwise all options can look good to the planners. The trick is to match options generated to that criteria so that the solution becomes fairly obvious vs. endless discussions and litigation based on differing opinions of the planners.

Generate multiple options

A true planning process should have multiple options to choose from. Even if the path is simple to decide, it must have opportunity costs—paths not taken—well articulated. The multiple options should be laid out side-by-side in a strategy-setting document with pros and cons listed for each. This allows planners to make an affirmative choice on different possible directions that can be taken and gain confidence by looking at and rejecting reasonable alternatives.

There is never one path to win in the history of product or business. It's always a question of the right paths. An added advantage of multiple options is that you have ideas to fall back on if the path chosen does not yield the desired result.

In rare cases, a company will have the time and money to try different paths in parallel. Think of it as a red team that prepares for alternate outcomes from the main plan and prepares a platform for more radical and different action. However, this is very rare and usually when done is lightly funded because of resource constraints.

Finalize the new direction

Ultimately, some direction needs to be picked and locked by the planning team. Ideally this is a principled direction selected through whittling the options based on a set of objective criteria. However, there is always room to debate in these situations and ultimately it is the CEO or CPO's job to anoint a specific plan and lay out the guard rails on how to attain it. This is usually done in some kind of strategy memo that lays out the direction for all staff. It can be accompanied with a meeting that lays out the process and the details of that direction. The goal is to orient the company around this new direction and get everyone convinced of its wisdom and fired up to go after its goals.

It's also worth noting that direction setting is not the same as an operational plan, which focuses on how Shipyard activities coordinate with the rest of the business, and ties into company financials. In many cases, when a direction is set, multiple people within the company are then activated to build a practical operational plan to attain the goals and metrics set in the strategic direction and to prosecute it to the best of their ability.

Sign off and get a commitment to execute it

Often, direction is disputed even among the planners, strategists and company leaders. Some of the people responsible for communicating and executing the new plan may still have reservations. This is understandable, but it's also critical that, at some point, these questions are set aside and the entire team commits to executing on the new direction.

One way to reduce the effect of this is to have a sign-off ritual of some sort. This can be part of communicating the direction as mentioned above. Or it can be a simple act of finalizing the strategy document and having each participant sign off or check off an acknowledgement of the new plan. This is usually referred to as a commitment device. Tech legend often calls this "disagree and commit"—a cultural practice usually attributed to Amazon. This practice means that while there may be dissent, for the foreseeable future, everyone agrees to work on the new plan and make it successful. The disagreements are meant to be aired, litigated, and exhausted during the planning process. It's an acknowledgement that planning is a team sport but that some authority gets to pick the final strategy proposal (usually a CEO or CPO) that goes into operation.

Staying on strategy and changing strategy when necessary are the real work of product leaders in technology companies today. Clear communication and decisive decision-making must be balanced with a sober view of reality and willingness to adjust when things aren't working as planned.

ALIGNMENT: KEEPING EVERYONE ON THE SAME PAGE

If the product Shipyard is controlled chaos, then *alignment* is the invisible thread that ties everything together. Without it, you risk teams working at cross-purposes, duplicating efforts, or worse, pulling your product in conflicting directions.

But here's the challenge: In a bustling Shipyard, it often feels like only those in specific areas know what's really happening. The key is to *make the invisible, visible*.

Systematic Information Collection

The foundation of alignment is systematic information collection. One affectionate term we use for this process is "product roundups." It's a structured way to gather intel from every corner of your Shipyard.

At our recent companies, we've made sure our teams implemented regular product roundups to ensure no critical information slipped through the cracks. Here's how you can do it:

1. Designate a point person in each team or department.

2. Have them contribute weekly updates on progress, challenges, and upcoming work. Think of this as team journaling.

3. Compile these updates into a digestible format, through an assembly process.

4. Share the compiled information widely, but thoughtfully.

If a Shipyard has a Product Operations specialist, it's a no-brainer responsibility for them to coordinate this. Remember, different audiences need different levels of detail. Your engineers don't need the same information as your executives or customers. Filter and tailor your communications accordingly.

Upward Flow of Information

It's not enough to collect information; you need to channel it upward to upper management in a disciplined way. This keeps the leadership teams informed and enables better decision-making at all levels.

We recommend creating a clear pathway for information to flow up the chain of command, even if it's not actively requested. This could be through regular reports, dashboards, or dedicated meetings where team leads brief partners, stakeholders and leaders.

Time spent designing this in such a way that the dissemination is automatically done is rarely wasted. Only small startups don't need this because they are hands-on all the time and information distribution is automatic.

Downward and Sideways Flow of Information

As mentioned, small teams and squads often have perfect information distribution (if they're doing it right). However, multiple squads or domains (such as mobile engineering) may not—coordination overhead is a real concern. Shipyards have to be mindful about information distribution once they get beyond a few teams. This helps each team make better decisions as they go.

Alignment should feel automatic and natural. Leaders accomplish this by making key direction update messages ambient across the organization. Anytime a key decision is made, question is asked, or priority is changed, the organization learns about it in real time. In addition, leaders must encourage a culture where it's ok to ask questions and gain clarification (without being obtuse and disruptive).

There are many tools to accomplish this—from Slack channels to regular emails and briefing meetings.

Aligment Rituals and Tools

To truly embed alignment in your Shipyard's DNA, establish systematic alignment rituals. Here are a few we've found particularly effective:

1. Demo days: Regular sessions where teams showcase their work. This isn't just about showing off; it's about cross-pollination of ideas and spotting potential synergies.

2. Firm agreement check-ins: When working with partner teams or dealing with dependencies, always end discussions with a clear articulation of what's been agreed upon. Document it and share it.

3. Decision logs: Maintain a centralized log of key decisions. This prevents rehashing old discussions and ensures everyone is on the same page about what's been decided and why.

While processes are crucial, the right tools can significantly enhance your alignment efforts. There are dedicated platforms designed to facilitate this kind of information sharing and alignment. For instance, Launch Notes is a tool specifically built for product communication in large organizations.

However, you don't need fancy tools to start. A well-structured document or internal wiki can be just as effective if used consistently. There are many more examples of great Shipyard rituals and tools! You probably have some good ones already. Check out the Coda team's collection of rituals and tools. The current CEO, Shishir, and his team have been compiling the best Shipyard rituals for a few years now and we know there is a book in the works.

The Benefits of Strong Alignment

When you get alignment right, the benefits are profound:

1. Reduced duplication: Teams avoid unknowingly duplicating work happening elsewhere.

2. Increased collaboration: When people know what others are working on, they can spot opportunities to join forces.

3. Faster decision-making: With everyone on the same page, decisions can be made more quickly and confidently.

4. Improved morale: Teams feel more connected to the bigger picture, boosting motivation and engagement.

Remember, alignment isn't about control or micromanagement. It's about creating a shared understanding that empowers teams to ask better questions, make better decisions, and work more effectively together.

In your role as a founder or product leader, fostering this alignment is one of your most crucial responsibilities. It's not always easy, and it requires constant attention, but the payoff in terms of productivity and product quality is immense.

Remember: Communicate Constantly

It's an axiom of good product leadership that at least half the job is communicating the vision, the mission, the plan, and the destination. The absolute importance of this is not an exaggeration, but many people reading this will still underestimate how critical this is. We will readily admit that we have failed multiple times because we did not grasp how important this skill and the execution of it was.

Product leaders have to very effectively communicate up, sideways, and down. We have found that doing this extensively and well pays huge dividends, especially if you have a large organization.

Starting with the board, at executive team meetings, at company all-hands meetings, and with each department product, leaders at every level should take every opportunity to address teams and key stakeholders about the destination they are rowing in and how they intend to reach it.

EPILOGUE

The Next Super Cycle:
Product Management for Artificial Intelligence

PRODUCT MANAGEMENT FOR AI

AI represents a new inflection point in building technology companies. Just as the advent of the internet transformed businesses in the late twentieth century, AI is now reshaping how we build and deliver software. This transformation is opening up unprecedented possibilities for startups and established companies alike. It's not merely an enhancement but a fundamental shift in approach, creating a new paradigm in product management.

However, we have to stress that AI will likely not change the fundamentals *this* time. Just like the internet, social, mobile, location, and the cloud did not change the fundamentals. Product managers and builders still have to discover sharp problems through customer discovery, refine their solutions via customer listening, design simple and powerful ways to solve customer problems, make it easy for them to try and buy them, etc. We have covered these fundamentals in Stage 1 and we strongly believe you will still need them in the AI age. Customer core needs and workflows are remarkably durable. What really changes is the level of technology and the amount of creativity we can bring to building solutions to them.

This epilogue explores the profound implications of AI on product management, from the emergence of transformer-based AI to the economic impact of AI-driven tools, and from the practical considerations of integrating AI into workflows to the new metrics that product managers must track. We will share our current thoughts on how AI is changing the landscape, the challenges and opportunities it presents, and the principles that will guide us through this transformation.

A BRIEF HISTORY OF AI

Artificial Intelligence (AI) has evolved significantly since its inception in the mid-twentieth century. Initially focused on symbolic reasoning and problem-solving, early AI efforts aimed to simulate human intelligence. However, progress was slow, leading to several AI winters—periods of reduced funding and interest due to unmet expectations.

The 1980s and 1990s saw a resurgence in AI, driven by advancements in machine learning (ML) and neural networks. The rise of deep learning in the early 2000s, leveraging convolutional and recurrent neural networks, enabled impressive breakthroughs in image and speech recognition. Despite these advancements, measurable and felt progress was slow. However, a significant leap occurred in 2017 with the introduction of transformer-based AI.

Transformer-based AI, introduced by Vaswani et al. in the 2017 paper "Attention is All You Need," revolutionized the AI landscape. Unlike previous approaches, transformers use self-attention mechanisms, allowing simultaneous data processing and focusing on the most relevant parts of input sequences. This innovation significantly enhanced the efficiency and accuracy of processing large datasets.

The impact of transformer-based AI is profound. These models excel in natural language processing (NLP) tasks such as language translation, text summarization, and sentiment analysis. Think of it as AI systems that understand human communication near perfectly (our input), can have a copy of the entire internet (which comprises a lot of human knowledge) in its memory as reference, and can create anything needed to communicate an accurate human-understandable output. Initially, these new AI models have been referred to as "large language models," or LLMs. However, given their recent multimedia capabilities in which they can ingest and synthesize text, images, and videos, we tend to refer to them as "large media models," or LMMs.

The scalability of transformer models allows them to handle vast datasets, providing deeper insights into customer behavior and market trends. This capability is essential for startups and established companies to innovate and compete globally. By leveraging transformer-based AI, companies can streamline decision-making processes and drive growth, paving the way for intelligent, AI-driven enterprises.

So what does that mean for product management?

A NEW TOOLBOX

Artificial intelligence is often described as just another (albeit powerful) tool. But it is more than that.

Rather than representing a single tool, we view AI as a new toolbox to solve workflow problems across organizations and across. AI-powered workflows are quickly becoming the norm as companies recognize the immense benefits of automation and intelligent data processing. From customer service and marketing to product development and logistics, AI enhances efficiency by automating routine tasks, analyzing data in real-time, and making predictive decisions. This shift allows human resources to focus on more strategic, creative, and value-added activities.

For example, in customer service, AI chatbots handle common inquiries, providing instant responses and freeing up human agents to tackle more complex issues. In marketing, AI algorithms analyze consumer behavior and predict trends, enabling more targeted and effective campaigns. In product development, AI-driven analytics identify potential areas of innovation and improvement, accelerating the development cycle.

One of the most profound changes AI brings is the democratization of advanced tools and capabilities. Previously, sophisticated AI systems were only accessible to large corporations with significant resources. Today, AI tools are becoming increasingly available to startups and smaller enterprises, leveling the playing field.

Cloud-based AI platforms and open-source frameworks have made it easier for companies of all sizes to integrate AI into their workflows. This democratization allows smaller companies to innovate rapidly, compete with larger firms, and bring new products to market faster. Startups can now leverage AI to analyze market trends, optimize operations, and create personalized customer experiences without needing massive upfront investments.

Time to Value is Significantly Reduced with AI

One of the most significant advantages of integrating AI into technology-enabled workflows is the reduction in time to value. AI accelerates processes that traditionally took hours or even days, delivering results in minutes. This rapid processing capability transforms the user experience and drives business agility.

Consider the process of analyzing large datasets. Previously, data scientists might spend days cleaning and analyzing data to extract insights. With AI, this process is expedited as machine learning algorithms and transformers can quickly process and analyze data, identifying patterns and trends with greater accuracy. This efficiency allows companies to make faster, data-driven decisions, enhancing their competitive edge.

This poses both a significant opportunity and challenge for established technology companies and startups alike. AI dramatically reduces the capital needed to build and scale a company. Traditionally, launching a startup required significant upfront investment in technology, infrastructure, and human resources. AI changes this paradigm by automating numerous tasks and processes, enabling startups to operate more efficiently with fewer resources.

AI-powered tools can handle everything from customer service and marketing to product development and data analysis. This automation reduces the need for large teams and extensive infrastructure, allowing startups to allocate their limited capital more strategically. For instance, AI can automate customer support through chatbots, manage social media campaigns with precision targeting, and streamline development cycles with predictive analytics. This means that a small team can achieve what previously required a much larger organization.

With lower capital requirements, barriers to entry are reduced, leading to a surge in innovation and entrepreneurship. Startups can now focus more on creative solutions and less on operational overhead. The democratization of AI tools ensures that even the smallest startups can compete with established companies, fostering a more dynamic and competitive market.

BUILDING WITH AI: TWO APPROACHES

AI offers two distinct approaches to integration within software: AI at the Core and AI at the Edge. Each approach has its own advantages and use cases, depending on the desired outcomes and the nature of the software.

AI at the Edge: Augment Existing Software Workflows

AI at the Edge involves integrating AI into already built software code and user experience workflows to enhance and optimize their performance. This approach augments the capabilities of traditional software by adding AI-driven features that improve efficiency, accuracy, and user experience. AI at the Edge is particularly effective for existing customer-facing applications, where it can significantly reduce the time and effort required to complete tasks.

Typeform is a prime example of AI at the Edge. Typeform has always been the most user-friendly and visually-appealing survey tool, but now it uses AI to enhance its functionality. The team worked on integrating AI-driven features into Typeform, such as intelligent suggestions for form questions, automated data analysis, and personalized customer interactions. These enhancements make it easier for customers to create engaging forms and surveys, analyze responses, and gain actionable insights. It used to take ten to thirty minutes to create a form. Now it takes one to three minutes.

This 10x improvement in productivity is currently AI's floor, but it is capable of so much more.

AI at the Core: Creating New Superpowers

AI at the Core involves building software where AI drives the fundamental assumptions and functionalities from the ground up. This approach embeds AI into the very essence of the solution being designed, making it integral to the software's operation and capabilities. When your new product or feature's code base is greater than 50% LMM, we tend to think of it as AI at the Core. The primary advantage of AI at the Core is that it enables the creation of applications that can perform complex tasks autonomously, introducing entirely new possibilities and capabilities.

Sticking with Typeform products, so we can contrast against the previous AI at the Edge example, let's look at Formless. Formless is one of the first examples of a product built with AI at the Core at Typeform. It uses AI to redefine how forms are created, filled, and processed. Instead of customers creating forms, Formless leverages AI to automatically generate forms based on customer input and behavior. The AI learns from interactions, continuously improving the form generation process to enhance user experience and efficiency. By integrating AI at its core, Formless offers customers a seamless, intelligent solution that simplifies data collection and processing.

AI at the Core does not just reduce toil and time to value. It introduces brand-new superpowers for customers. Applications built this way can anticipate customer needs, automate complex workflows, and provide insights that would be impossible to achieve with traditional software.

Both AI at the Core and AI at the Edge paradigms offer powerful ways to integrate AI into product development. The choice between the two approaches depends on the desired outcomes and the the target market and its customers and the problem space. AI at the Core can completely revolutionize how software operates, while AI at the Edge can enhance and optimize existing workflows, delivering significant benefits to customers and businesses alike. AI at the Core often requires a rethinking of the problem space and can take time or be unnerving for organizations. AI at the Edge is often more immediately accessiblle to established companies since it often isn't as disruptive to established patterns.

However you choose to build with AI, it will require new ways to measure success.

AI METRICS: MEASURING SUCCESS

As you embark on AI-powered product solutions, we have some recommendations on how to update your existing measurements of success and feature metrics with such investments. We recommend starting with these four customer-centered metrics.

Time to Value

Time to value measures the speed at which an AI solution delivers tangible benefits to the customer or business. This metric is crucial because one of AI's primary advantages is its ability to accelerate processes and reduce the time required to achieve results. A shorter time to value indicates a more effective and impactful AI implementation. For instance, if an AI tool can analyze data and provide insights within minutes, as opposed to hours or days, it significantly enhances operational efficiency and decision-making.

Customer Return Rate

Many customers will be curious about your new AI products or features, but how many return to use it again? We have personally tried many AI tools just once and never returned. Usage Return Rate (URR) helps determine the value and relevance of AI capabilities to customers. High URR suggests that the AI features are meeting customer needs and providing substantial benefits.

Conversely, low URR may indicate that the features are not well-integrated, not user-friendly, or not perceived as valuable. Regularly monitoring AI feature usage can guide product improvements and ensure that AI enhancements align with customer expectations and needs.

Workflow Length

Workflow length measures the duration of specific tasks or processes before and after AI integration. By comparing these durations, businesses can quantify the efficiency gains provided by AI. A significant reduction in workflow length demonstrates that AI is effectively streamlining operations and reducing the effort required to complete tasks. For example, if a data analysis process that previously took several hours can be completed in minutes with AI, the reduction in workflow length is a clear indicator of AI's value.

Tweak Time

Tweak time refers to the amount of time customers spend refining and adjusting AI-generated outputs. While AI can automate many tasks, it often requires human oversight to ensure accuracy and relevance. Tracking tweak time helps identify areas where AI models need improvement and where additional training data or algorithm adjustments are necessary. Ideally, as AI systems become more advanced and better trained, tweak time should decrease, indicating that the AI is producing more accurate and reliable results with less human intervention.

Measuring success in AI-driven projects requires new metrics that reflect the unique capabilities and benefits of AI. Time to value, AI feature usage, workflow length, and tweak time provide a comprehensive framework for assessing the impact of AI implementations. By focusing on these metrics, businesses can ensure that their AI investments deliver meaningful and measurable improvements, driving innovation and competitive advantage in the AI age.

Building Rocketships in the Age of AI and Beyond

Artificial Intelligence does not change the rules of product management. It does not remove the need for customer centricity, deep understanding of customer needs, or the importance of delivering meaningful solutions.

What AI does is open the door to rethink *everything* around us. No entrenched product, platform, or workflow is safe from disruption. If you can dream it, you can build it—and build it faster and cheaper than ever before. In a sense, AI is the dawning of the Product Golden Age, where technological limitations nor incumbents can stop the best ideas from winning. There has never been a more exciting time to work in technology.

We started this book with a simple question: How are rocketship companies built? Now we hope you have a few of the answers you need and perhaps some inspiration as well.

The AI revolution is unleashing a Cambrian explosion of innovation and experimentation. While much remains unknown, it's clear that AI represents a step change in how we build and deliver technology-driven solutions to customer problems.

Forward-thinking builders must embrace the uncertainty and dive in—identifying sharp use cases where AI can have an outsized impact, developing a point of view on AI's role in their domain, and cultivating the judgment to implement AI responsibly and effectively. The future belongs to those who can harness the power of AI to solve customer problems in truly transformative ways.

We are rooting for you. Let's build!

ABOUT THE AUTHORS

Oji Udezue is a former Chief Product Officer, investor, and entrepreneur. He has led Product, Engineering and Marketing at beloved companies like Microsoft, Twitter, Calendly, Atlassian, Typeform, and more. Today Oji is pushing the edges of building AI products by working with established companies and investing in next-generation startups and their founders.

Ezinne Udezue is a former Chief Product Officer, investor and expert in digital transformations. She had led enterprise innovation initiatives at T-Mobile, Discovery Communications, Time Inc., Procore, and WP Engine. Ezinne has demonstrated expertise in scaling product-led organizations.

They work with and advise select clients at

PRODUCTMIND.CO

www.ingramcontent.com/pod-product-compliance
Lightning Source LLC
Chambersburg PA
CBHW030455210326
41597CB00013B/680